Marketplaces

This edited volume portrays marketplaces from a mobility perspective as dynamic and open entities consisting of flows of people, goods and ideas.

There is a renewed interest in research and policy arenas in marketplaces as the core of cities' spatial and economic development and sociocultural life, as incubators of urban renewal and platforms of alternative consumption models and as a source of livelihood for many people worldwide. Contributions of this book draw on notions of movements, representations and practices to illustrate that markets have physical reality but are also culturally and socially encoded, and experienced through practice. It brings together empirically evidenced scholarly and practice-based works from the United Kingdom, the Netherlands, Switzerland, Spain, Bulgaria, Turkey, Lebanon, Peru, Brazil, Vietnam, South Africa and India.

This book is primarily intended for scholars and graduate students of urban geography, urban design and planning, sociology, anthropology, who are interested in the relation between place and mobility in general, and markets as 'knots' in the city, in particular. It also informs policy-makers how urban planning policies and design interventions for marketplaces may foster more socially inclusive and environmentally just cities.

Ceren Sezer is a Research Associate at the Institute for Urban Design and European Urbanism of Aachen University, Germany. She is joint editor of *Marketplaces as an Urban Development Strategy* (2013), *Public Space and Urban Justice* (2017) and the author of *Visibility, Democratic Public Space and Socially Inclusive Cities* (2020).

Rianne van Melik is an Associate Professor in Urban Geography at the Institute of Management Research at Radboud University Nijmegen, the Netherlands. Her research focuses on contemporary cities and their public spaces, with specific interests in the design, management, use and perception of different kinds of public spaces. She is principal investigator of the Moving Marketplaces (MMP) project.

Routledge Studies in Urbanism and the City

For more information about this series, please visit https://www.routledge.com/
Routledge-Studies-in-Urbanism-and-the-City/book-series/RSUC

Marketplaces

Movements, Representations and Practices

Edited by Ceren Sezer and
Rianne van Melik

Routledge
Taylor & Francis Group

LONDON AND NEW YORK

First published 2023
by Routledge
4 Park Square, Milton Park, Abingdon, Oxon OX14 4RN

and by Routledge
605 Third Avenue, New York, NY 10158

Routledge is an imprint of the Taylor & Francis Group, an informa business

British Library Cataloguing-in-Publication Data
A catalogue record for this book is available from the British Library

Library of Congress Cataloging-in-Publication Data
A catalog record has been requested for this book

ISBN: 978-1-032-05325-7 (hbk)
ISBN: 978-1-032-05326-4 (pbk)
ISBN: 978-1-003-19705-8 (ebk)

DOI: 10.4324/9781003197058

Typeset in Goudy
by SPi Technologies India Pvt Ltd (Straive)

For Deniz and Nurcemal Nelissen
&
For Kris De Prins

Contents

Illustrations

Figures

Tables

Contributors

Markus Breines is an Assistant Professor in Social Science at the London School of Hygiene and Tropical Medicine, the United Kingdom. He studies the relationship between migration and social mobility. He was part of the international research project Moving Marketplaces (MMP): Following the Everyday Production of Inclusive Public Space, funded by Humanities in the European Research Area.

Bilge Beril Kapusuz-Balcı is a Research Assistant at Gazi University Architecture Department in Ankara, Turkey. Her PhD dissertation focused on the historical archives of the Venice Biennale. Her research includes basic design, art and architectural theory and criticism, visual culture in relation to architecture and urban design, focusing on exhibition phenomena and exhibition histories on a transdisciplinary ground.

Janine Dahinden is a Professor of Transnational Studies and Director of the MAPS (Maison d'analyse des processus sociaux) at the University of Neuchâtel, Switzerland. She investigates processes of mobility, transnationalisation and boundary making, and their concomitant production of inequalities linked to ethnicity, race, class, religion or gender. She is principal investigator of the MMP project.

Sara González is Associate Professor at the School of Geography of the University of Leeds, the United Kingdom. She has been involved in academic and action research on markets for over ten years. She is interested in cities as spaces of contestation and how value is created in urban space.

Ana María Huaita Alfaro holds a PhD in Development Planning by the Bartlett Development Planning Unit, University College London, the United Kingdom. She currently works as a specialist on urban food markets and food systems for FAO Investment Centre in Rome, Italy. Her recent work focuses on urban marketplaces and encounters around food and negotiations for everyday urban living.

Kiran Keswani is co-founder of Everyday City Lab, an urban design and research collaborative in Bangalore, India. She has a PhD from the Faculty of Planning, the Centre for Environmental Planning and Technology University,

Ahmedabad. She focuses on informality in public spaces and everyday practices of people.

Karina Landman is an Associate Professor in the Department of Town and Regional Planning at the University of Pretoria, South Africa. Her current research interests relate to public space, nodal development and urban resilience and regeneration.

Maria Lindmäe is a postdoctoral researcher at the Department of Humanities, Universitat Pompeu Fabra in Barcelona, Spain. As a cultural geographer, she is interested in public space, urban transformations and music urbanism. She currently studies street markets as part of the MMP project.

Marco Madella is Catalan Institution for Research and Advanced Studies Research Professor in Environmental Archaeology at the Department of Humanities, Universitat Pompeu Fabra in Barcelona, Spain. He is interested in food production and early urbanism. He is principal investigator of the MMP project.

Christine Mady is an Associate Professor in Urban Planning, Urban Design and Architecture at Ramez G. Chagoury Faculty of Architecture, Arts and Design at the Notre Dame University-Louaize in Lebanon. Her research interests include public space and social integration in unstable contexts, temporary urban uses, heritage landscapes, urban planning and public health and planning for social infrastructure.

Nihan Oya Memlük-Çobanoğlu is a Research Assistant at the City and Regional Planning Department at Gazi University in Ankara, Turkey. She received her doctorate degree from the City and Regional Planning Department at Middle East Technical University. Her research fields include basic design, public space, urban design and environmental psychology.

Joanna Menet is a postdoctoral researcher at the University of Neuchâtel, Switzerland. She works at the intersections of scholarship on migration/mobility, ethnicity, gender and dance. She currently studies street markets as part of the MMP project.

Noxolo Ndaba is a PhD candidate in the Department of Town and Regional Planning at the University of Pretoria, South Africa. Her PhD research focuses on the street as a public space to generate opportunities for sustainable livelihoods. She is also currently employed by Jozini Municipality as an Executive Director: Planning and Economic Development.

Joris Schapendonk is an Associate Professor in Human Geography at the Institute of Management Research at Radboud University Nijmegen, the Netherlands. He focuses on migrant trajectories, undocumented migrants and mobility practices. He is project member of the MMP project.

Patrícia Schappo is a postgraduate researcher at the Department of Urban Studies and Planning of the University of Sheffield, United Kingdom. Her research interests relate to planning, diversity and integration and the attainment of justice at the urban level.

Sarah Turner is a Professor in the Department of Geography, McGill University, Canada. Her research focuses on informal economy livelihoods of Hanoi and smaller cities in Vietnam's uplands. She completed similar previous work in Indonesia. She also studies ethnic minority livelihoods in the Sino-Vietnamese borderlands.

Emil van Eck is a PhD student at the Institute of Management Research at Radboud University Nijmegen, the Netherlands. He currently studies street markets as part of the MMP project.

Nikola A. Venkov is an Assistant Professor at the Institute of Philosophy and Sociology at the Bulgarian Academy of Sciences. His interests are in the field of urban anthropology, studying power and marginalisation in urban space and politics. His work on these topics also expands over to public and engaged anthropology and even socially engaged art (he is a co-founder of Duvar Kolektiv).

Sophie Watson is a Professor of Sociology at the Faculty of Arts & Social Sciences of the Open University in Milton Keynes, United Kingdom. Her research revolves around marketplaces, city water and spatial justice. She is project leader of the MMP project.

Celia Zuberec is an honours student in the Department of Geography, McGill University, Canada. Her honours research focused on youth itinerant street vendors in Hanoi, Vietnam. She has interests in rural-to-urban migration, informal livelihoods, public space and the creative economy.

Acknowledgements

The editors are grateful to all those who accepted our invitation to take part in this book. We are particularly thankful to the Humanities in the European Research Area Joint Research Programme "Public Spaces: Culture and Integration in Europe", who financially supported the research project *Moving Marketplaces (MMP): Following the Everyday Production of Inclusive Public Space*. The project's mobility perspective forms the foundation of this book, and consortium members authored five chapters. The RWTH Aachen University Institute and Chair for Urban Design led by Professor Christa Reicher generously supported Ceren Sezer when she was awarded the Deutscher Akademischer Austauschdienst grant. In addition, we would like to thank Routledge, specifically our editor Faye Leerink for her support and guidance during the preparation of the book. We are also grateful to Professor Sara González for accepting our invitation to write an afterword for our book.

Finally, we thank our families for their patience and endless support during our book's editing process, especially in the challenging times of COVID-19 lockdowns. This book is dedicated to them.

Ceren Sezer
Rianne van Melik
December 2021

1 Introduction

Rianne van Melik and Ceren Sezer

Introduction

This edited book investigates everyday marketplaces as important socio-economic spaces through a mobility lens. On the one hand, markets are often seen as quintessential public *places*, where people of different social, economic and cultural backgrounds feel welcome and can interact (Morales, 2009; Watson, 2009; Janssens & Sezer, 2013a). On the other hand, markets are perhaps the most *mobile*, dynamic, temporal and translocal of all public spaces (Seale, 2016; Schappo & Van Melik, 2017; Breines et al., 2021), constructed and deconstructed each operating day and consisting of a wide array of people, goods and ideas. Hence, markets provide a unique opportunity to study the relation between place and mobility.

In social sciences, the mobility perspective views social and cultural changes not as static processes but as various states of spatial and temporal transformations produced by the movement of human and non-humans, including things, objects, ideas and information (Sheller & Urry, 2006; Urry, 2007; Adey, 2017). In a similar vein, this book presents empirically evidenced research which studies marketplaces as dynamic and open entities rather than primarily as bounded places where economic transactions occur and encounters take place (Watson, 2009; Hiebert et al., 2015). It brings together various scholarly and practice-based works on marketplaces, traders, visitors, regularity bodies, designers and planners from a wide geographical distribution, including the United Kingdom, the Netherlands, Switzerland, Spain, Bulgaria, Turkey, Lebanon, Peru, Brazil, Vietnam and South Africa.

Such a relational approach is not new in market research; previous studies on markets, including earlier Routledge books, have also focused on markets as nodes of flows (e.g. Seale, 2016; González, 2018). However, what is new in this book is that we make this relational perspective more explicit by combining market studies with conceptual starting points from mobilities studies. Building upon the work of Tim Cresswell (2010), among others, we look at movements, representations and practices on/of the markets to illustrate that their importance often stretches beyond their physical confines (Schappo & Van Melik, 2017).

Our purpose in this collection of work is threefold. First, with increasing attention on marketplaces in research, practice and policy arenas, we believe

DOI: 10.4324/9781003197058-1

that a mobility perspective can bring a renewed approach to marketplace studies. Second, we aim to stimulate better communication between marketplace researchers and practitioners from various disciplines, such as urban geography, sociology, anthropology, urban design and planning and policy-making. A mobility perspective can serve as a helpful framework that ties together and generates dialogue between these various fields of marketplace studies. Third, we aim to present that marketplace may potentially crosscut geographical and cultural boundaries, blurring the Global North and Global South divide as they are very often addressed in social science studies. The mobility perspective provides a useful framework to present common and distinctive features of marketplaces beyond their geographical contextualisation.

The following section conceptualises marketplaces and explains why and how they gained increasing attention from researchers and policymakers throughout history. The chapter then elaborates how a mobility perspective helps to extend this state-of-the-art and to foster increased multidisciplinary dialogue on marketplaces. After presenting the book's outline, the chapter ends with some reflections on the development of this edited volume during a worldwide pandemic.

Conceptualisations of marketplaces

Marketplaces have long attracted research attention due to their essential role as the heart of urban settlements. In the Greek agora, marketplaces were sites of daily improvisational performance (Sennett, 1998). The display of goods, fruits, flowers, contrasting colours and scents, together with the vendors' cheerful voices, served as a recognition space where social interactions were performed without any specific or scripted roles. This colourful landscape of the marketplace inspired Agnew's book – *Worlds Apart, The Market and the Theatre in Anglo-American Thought* (1986), in which he compares marketplaces to theatres, another spatial and cultural stage. Agnew argues that, unlike theatres, marketplaces break all kinds of social and physical limitations with their multivalued, random and spatial relations. Accordingly, a marketplace functions as the threshold of exchange between different social worlds (Agnew, 1986).

Marketplaces were also generally seen as the primary driver of urban growth. Historically, the main source of city development was trade, and the marketplaces were at the core of these exchanges in which various goods were imported and exported (Jacobs, 1969; Stobart & Van Damme, 2016). The city centre of market towns was essentially devoted to the market, which was regularly settled on certain days (Kostof, 1999). The location of the marketplace was chosen in an easy access to the city's entry points and main trade roads. For example, marketplaces in 12th and 13th century England were mainly seen at the border between the villages. The marketplace was positioned as a neutral zone where people met regularly for commercial transactions (Kelley, 2016).

Marketplaces are also understood as a component of the cultural heritage, as they fundamentally link to the understanding of time, space, local identity, cultural practices and spirituality. A significant example of this is Nigeria's ethnic community, the Igbos, who see the world as a marketplace where individuals are born to be traders and die to go back for the next cycle of reincarnation (Chukwuemeka et al., 2020). The Igbos believe that societies without trade or other forms of cultural and material exchanges are subject to conflict and contestation. The market days are also the basis for their calendar system in which a week consists of four days and seven weeks make a month. Even after adopting the Gregorian calendar system, market days still play a role in the Igbos' lives to decide farming periods, days for marriage ceremonies and for burial rites. The Igbos are still prominent merchant communities in Nigeria (Chukwuemeka et al., 2020).

Despite the significant importance of marketplaces as the core of economic and socio-cultural transactions in the city, local authorities often tend to problematise them as unhygienic and unhealthy urban environments. Early examples of this situation were in the major European cities in the mid-19th century, where poverty, overpopulation and pollution were the main problems in inner-city areas. For example, in London, street markets were a part of the vivid urban scene in the 19th century supplying cheap food and products (Mayhew, 1985). However, they were unorganised and naturally growing (Kelley, 2016). The city authorities have viewed these markets as components of the city's degraded living conditions. They introduced structural spatial changes to address this problem, including removing street markets and developing new and enlarged indoor marketplaces. These new indoor markets functioned as an urban renewal tool as well, as their construction required the demolition of existing building blocks and re-organisation of streets (Hall, 1988). They were also strictly organised with new tax rules, price control, opening and closing times and reflecting lifestyles of the affluent groups through their characteristic architecture and improved hygienic conditions (Kelley, 2016).

Today, marketplaces are still seen as an urban renewal tool in many cities worldwide. The city authorities often characterise them as problem areas due to health, safety and traffic concerns and illegal practices and tend to re-organise and brand them following various urban development agendas (Van Eck, 2021). They have been closed down or/and relocated in places to make them suitable for the modernisation agenda of the city authorities, such as the Kigali market in Rwanda (Michelon, 2009). Marketplaces also facilitate urban rebranding and place-making practices, such as Quincy Market in Boston and South Street Seaport in New York City (Janssens & Sezer, 2013a). They have been seen as tools for the commodification of immigrant cultural economies, such as Vancouver's Chinatown Night Market (Pottie-Sherman, 2013) and Amsterdam's Dappermarkt (Janssens & Sezer, 2013b). They have also been part of gentrification processes in the inner-city areas subject to market pressures, as seen in Cusco's marketplaces in Peru (Seligman & Guevara, 2013) and Beijing's Silk Street Market (Lin Pang & Sterling, 2013).

Nevertheless, a steadily growing body of literature reminds us of what marketplaces can offer cities and their people. As flexible spatial-temporal organisations, marketplaces can facilitate a spontaneous synergy between people of different social-economic and cultural backgrounds (Morales, 2009; Watson, 2009; Sezer, 2020). They can give a sense of the city's life and 'soul' (Urbact Markets, 2015). Markets worldwide share certain commonalities, which makes them familiar environments with similar routines and codes of conduct (Ünlü-Yücesoy, 2013). Along with relatively low entry barriers for traders, marketplaces are generally thought of as inclusive spaces, where diverse people feel they have an equal right to be (Morales, 2011; Hiebert et al., 2015; Nikšič & Sezer, 2017).

Recent studies have also shown that marketplaces can provide platforms for alternative consumption models of sharing and upcycling, reuse and recycling at non-monetary-based private and public sharing events. For example, Albinsson and Perera's (2012) research on 'Really Free Markets', initially organised by the Anarchist Movement, developed as a community movement including participants from wider groups and organisations.

And perhaps most importantly, the marketplace constitutes part of the minimal urban infrastructure required to meet basic human survival needs in many cities worldwide (Abwe, 2020). They function as the economic engine that connects rural farmers to urban traders and urban consumers, as an active part of the supply and demand chain as the basics of economics. They serve as grounds for job opportunities, livelihood sources, touristic attractions, education platforms for entrepreneurial skills and growth of wealth (Asante & Helbrecht, 2020). We, therefore, regard marketplaces as essential infrastructures which deserve wider recognition and deepened understanding, which this book aims to provide by applying a mobility perspective.

Marketplaces and mobility perspective

Although the plethora of research described earlier offers excellent accounts of the importance of marketplaces, we believe most of these studies so far have been relatively 'place-based' by focusing on what happens within the physical confines of the market: the encounters, transactions and activities occurring in the marketplace. Often, the market's design or the everyday interaction on the market is studied, showing how they are experienced and consumed, and function as meeting grounds supporting inclusive city life (e.g. Watson 2009; Polyák, 2014). An excellent example is De La Pradella's (2006) book entitled *Market Day in Provence*, in which she describes the social world of a weekly market. Another one is Marovelli's (2014) work on sensory perception of a Sicilian urban marketplace, which illustrates how certain smells and sounds at the market are allowed while others are not. Such place-based ethnographies approach markets as 'bounded containers of value' (Massey, 2001: 16): static, closed entities with clear boundaries in space and time, and their specific spatial design, organisational structure, economics and sociability.

While such studies greatly enhance our understanding of particular spaces, they provide little knowledge on their broader purpose beyond their physical and organisational boundaries or the relations between different public spaces. Conversely, Schappo and Van Melik (2017) show that the integrative potentiality of markets reaches beyond their physical confines, bringing together municipalities, traders, local entrepreneurs and residents. In other words, a focus on marketplaces as 'bounded containers' leads to limited insights regarding the question of how these public spaces are created by the constant incoming and outgoing flows of people, goods and ideas. Following Van Melik and Spierings (2020), this book, therefore, aims to go beyond 'place-based' studies and applies a more 'process-oriented' or 'processual' investigation that approaches markets as dynamic, fluid and open entities. As such, we aim to further extend our understandings of marketplaces by studying not just what they are but also how they *come into being*.

To develop such a processual understanding of public space, we draw for inspiration on the so-called new mobilities paradigm or mobility turn (Sheller & Urry, 2006), which resulted in a large and still growing number of studies focusing on the flows and movement of people, goods and ideas in social sciences (e.g. Adey, 2006, 2017; Urry, 2007; Jensen, 2009; Cresswell, 2010, 2011; Breines et al., 2021). While this mobility perspective is very common in certain fields of urban research, such as migration studies, it is less conventional in public space research in general, and market research in particular (with some notable exceptions, such as Qiang, 2013). Only street vending studies (e.g. Etzold, 2014; Neethi et al., 2021) tend to have a strong focus on flows and movements, such as Endres and Leshkowich (2018) edited volume *Traders in Motion: Identities and Contestations in the Vietnamese Marketplace*. This book describes the dynamics of emplacement, mobility and boundary-making practices of Vietnamese street vendors. Although the editors claim that the findings transgress Vietnam and also offer insights into other rapidly shifting trade contexts elsewhere in the Global South, we argue that a mobility perspective is also valuable in other spatial contexts.

Such a process-oriented, mobility perspective resonates with the increasing body of literature in urban studies that uses a relational lens to understand urban phenomena. Though the terminology differs from practices and assemblages to interplay, constellations or throwntogetherness (e.g. Massey, 2005; Cresswell, 2010), these studies share a common ground in looking at socio-material relations. Following this idea, it is not merely the (inter)actions of traders and consumers or the physical and legal structure of the market (i.e., design, regulation), but the interweaving of all these factors that determines how marketplaces function. As such, markets should quite literally be seen as intersections (Hiebert et al., 2015) where people, products and policies temporarily cross paths.

Figure 1.1 clearly illustrates these two different approaches to studying marketplaces. Inspired by the work of anthropologist Tim Ingold, Breines et al. (2021, 9) set off a place-based, bounded version of the market on the left against a mobile perspective on marketplaces on the right, in which they are perceived as 'knots' of people, things, goods and ideas. Although the version of the left also consists of straight lines indicating the routes to the market, "the lines represent statics movements and do not shape and transform the places but leave them untouched" (Breines et al., 2021: 10). In contrast, the market on the right only exists by merit of the mobilities of traders, customers, goods and stalls, without which there would not even have been a market. Mobility and public places thus not only exist simultaneously; mobilities produce places (Breines et al., 2021: 9).

To structure our argumentation, we draw on the work of human geographer Tim Cresswell in this edited volume. He defines the mobility perspective in social sciences as the exploration of "the movement of people, things, and ideas as at

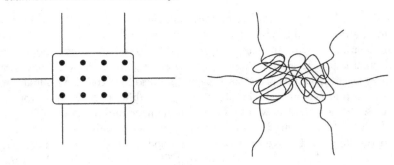

Figure 1.1 Two versions of a marketplace.
Source: Breines et al., 2021: 9.

the centre of *constellation* of power, the creation of identities and the micro geographies of everyday life" (Cresswell, 2011: 551). Following Walter Benjamin, Cresswell (2010: 18) uses the notion of 'constellations' to describe how mobility can be understood as a combination of (1) particular patterns of physical *movement*, (2) *representations* of movement, and (3) *practices* of movement, which only make sense together. In this edited book, we use a similar framework to study marketplaces.

According to Cresswell (2010: 19), *patterns of movement* are about getting from one place to another; this physical movement is "the raw material of the production of mobility. People move, things move, ideas move". When applied to marketplaces, this physical movement is about their mobile character, notably of street markets and street vending practices. Due to their ambulant nature, many markets move from one place to another. Being in a different locality each day, the traders have to cope with different institutional arrangements, markets' composition, layout and consumers. Often, the market does not move as a whole, but as separate entities. Such changes require flexibility from the traders, as well as navigational skills. However, not only traders move; markets are inherently spaces of flows – also of goods and ideas. They are a temporary get-together, which turns physical spaces in the city into public spaces – if only for the duration of market hours, after which they are disentangled again.

However, according to Cresswell, understanding physical movement is only one aspect of mobility: "But this says next to nothing about what these mobilities are made to mean or how they are practiced" (2010: 19). He therefore also focuses on different meanings – or *representations* – attributed to movement. Representations touch upon dominant narratives about the meaning and value of markets, also known as 'place frames' (Martin, 2003). While in some cases, markets are perceived as 'relics of the past' on the verge of distinction, there are also dominant discourses that claim that markets still (and increasingly so) have important socio-economic values in providing among others income, access to fresh food, conviviality and care, as outlined previously. Such places frames are not neutral; they are constructed, contested and negotiated. Van Eck (2021), for example, illustrates how the frame of what constitutes a 'good quality market' in Amsterdam is very much informed by dominant racialised ideas, which in turn influence the ways market managers, visitors and traders interact with each other.

Lastly, *practices* are everyday activities and how these are experienced and embodied. Cresswell (2010) mentions walking and driving as mobility practices, which are enacted and experienced through the body. In a similar vein, we focus on the different doings being performed at marketplaces—from planners and designers working on mitigating tensions to (ethnic) entrepreneurs who act as brokers through their bridging role between cultures and classes. Some of these practices are very embodied: the physical labour of trading, for example, requires many skills and strength.

Applying a mobility perspective to study marketplaces means that we think of them as entanglements of movement, representation and practice. Cresswell (2010) uses the example of driving to illustrate the value of such a 'constellation' or entanglement: to understand driving, one should not only look at the physical movement of getting from one place to another by car but also appreciate that this is imbued with certain narratives or values (freedom, polluting) and felt by the body (liberating, tiring, nerve recking). Cresswell (2010: 20) concludes, "Similar sets of observations could be made about all forms of mobility: they have a physical reality, they are encoded culturally and socially, and they are experienced through practice". In a similar vein, this book uses these three themes to enhance our understanding of marketplaces as dynamic, networked and multi-layered spaces.

Outline of the book

This volume contains 14 chapters, including this introductory chapter, which sets the ground for the argument of this book, and an afterword written by Sara González. The chapters all apply a mobility perspective to investigate marketplaces in different parts of the world. While some chapters explicitly focus on movement, others address certain dominant representations of markets, and almost all discuss certain (trading, policy) practices in one way or another. As Cresswell (2010: 19) also acknowledges, these three elements are difficult to disentangle, as they are bound up with one another. Therefore, we decided not to divide this book into clearly separated parts but to cluster chapters with similar foci.

Before we briefly introduce the chapters, it is necessary to note that we do not predefine marketplaces. Endres and Leshkowich (2018: 2) indicate that "the market is not a singular entity". Due to their diverse modalities, markets come in many different types and sizes. Our call for contributions (see the following) attracted scholars studying a wide range of places and activities, from street vending in Vietnam (Chapter 2) and India (Chapter 3); to indoor or covered markets with permanent infrastructures in Spain (Chapter 6), Brazil (Chapter 9) and Perú (Chapter 10); day-markets in Turkey (Chapter 4) and the UK (Chapter 11); and open-air farmer markets in Switzerland (Chapter 12). We appreciated the diversity of markets discussed in these contributions, as they show us that markets are indeed not singular—but are very different in time and space. If anything, we can only say what markets are *not*: they are *not* just bounded spatial locations.

We invited our contributors to empirically investigate not just the 'here-and-now' of their cases (for example, focusing on their use or design) but also the 'there-and-then', making links to other places or times. This approach resulted in researchers employing 'go-along' methods, among others, acknowledging the flexible and itinerant nature of trading, such as following research participants during their trading activities (Chapters 2 and 3). Other contributors have not followed traders but traced back where market legislation or market goods come from (Chapters 12 and 13). These chapters show how markets are often portrayed as local entities, yet are part and parcel of global networks of policies and goods. Another group of contributors has taken the temporal dynamics of markets into account, showing the changing rhythms of the market during the day or even the decade (Chapters 3, 4 and 8). Whatever their specific methodology, these contributions all acknowledge markets as unbounded, fluid entities.

Chapter 2 by Celia Zuberec and Sarah Turner focuses on vending strategies and mobility patterns of street vendors in Hanoi, Vietnam. Drawing on interviews with 35 migrant traders, including 'walking (or riding) while talking interviews', they show how mobility is differentially accessed, experienced and embodied. The traders draw on a range of dynamic strategies and everyday practices to reach potential customers while avoiding state sanctions simultaneously. Vendors exactly know where and when (not) to trade, and inform each other through social networks. Different narrative maps illustrate that being on the move—while perhaps suggesting flexibility and irregularity—can nevertheless result in dominant routes and rhythms.

In Chapter 3, Kiran Keswani uses the notion of rhythmanalysis, drawing on the work of philosopher and sociologist Henri Lefebvre. Her study of Bhadra Plaza in Ahmedabad in India portrays the (im)mobility patterns of push-cart vendors selling chai tea and their 'runners' delivering the kettles and cups to customers. Biological and social rhythms influence the movement of the chai runners; thirst and work breaks result in peaks of flows of people (runners) and goods (tea). Hence, the chapter illustrates the spatial and temporal relation between social interaction, time and movement in the marketplace.

In a similar vein, Nihan Oya Memlük-Çobanoğlu and Bilge Beril Kapusuz-Balcı visualise a day in the Esat Marketplace in Ankara (Turkey) in Chapter 4. In their spectral analysis, the authors show that there is a variety of rhythms and actors/actants producing them, including regular patterns (*eurythmia*), as well as moments of contingent rhythms (*arrhythmia*). Taking Lefebvre's rhythmanalysis as a starting point, they argue that we should not just study marketplaces as important urban spaces but also as timespaces or rhythmic fields.

Noxolo Ndaba and Karina Landman study the adaptive capacity of street traders in one of South Africa's largest transport hubs, Warwick Junction in eThekwini Municipality. Their findings in Chapter 5 illustrate how traders are resilient in response to changes like urban renewal projects and the COVID-19 pandemic. This is demonstrated in practices such as self-organisation and collaboration, and being mobile in search of the best (legal) trading location.

Urban renewal is a central theme in many chapters. The role of markets in such a renewal process can be various. Maria Lindmäe and Marco Madella focus in Chapter 6 on La Boqueria in Barcelona (Spain), which is often represented as the 'best' or 'most important' market in the world, a success story of market management and urban restructuring triggering policy tourism and international knowledge exchange. However, COVID-19 has illustrated how dependent the market has become on tourists and more affluent consumers seeking an interesting food experience rather than being a neighbourhood market offering a large variety of products with accessible prices. The current aim is to bring back the 'traditional' market. Still, public and private interests clash as the public administration, market managers and stallholders have different interpretations of what and for whom the market should be.

In other contexts, more negative representations of the market prevail. In Chapter 7, Christine Mady studies the Souq Al Ahad in Beirut, Lebanon. She shows how the market is in constant, incremental transformation and that different (political) stakeholders depict it as a place of disgrace and destitution, arguing for its relocation. Mady employs the concepts of liminality and fluidity to explain the market's role amidst these contestations and changes.

Similarly, Nikola A. Venkov interrogates in Chapter 8 what neoliberal restructuring looks like at two open-air markets in the post-socialist context of Sofia, Bulgaria. He uses the notion of socio-spatial sorting to describe how urban policy changes, such as the introduction of a new auction system stimulating intense intra-marketplace competition, resulted in a process of sorting both traders and buyers: "Poorer citizens were 'sorted towards' the Women's Market, while food growers were 'sorted out' of marketplaces."

Such patterns of contestation and competition indeed raise questions regarding who the market is for and to what extent it serves as a *just* place. In Chapter 9, Patrícia Schappo investigates the ongoing transfer of responsibility for market management away from the public to the private sector at four public markets in Belo Horizonte, Brazil. She evaluates a recent tendering process by drawing upon Fainstein's justice triad of democracy, equity and diversity. Rather than labelling such shifting responsibilities as neoliberal strategies, she concludes that markets are infrastructures where entrepreneurial *and* social justice aims can coexist.

All these contributions touch upon certain dominant narratives that exist about the meaning and value of markets. Ana María Huaita Alfaro discusses in Chapter 10 public expressions regarding marketplaces in Lima, Peru. In contrast to other (Latin American) cities, public markets here still play an essential role in food provision, growing in number and outperforming supermarkets. However, the chapter shows that the appreciation of these markets beyond their function as commercial spaces is limited. Markets are generally represented as neglected, outdated and non-modern, which restricts their potential function as common ground for commingling.

Chapter 11 by Sophie Watson and Markus Breines discusses another dominant representation regarding the relationship between traders' mobility and their attachment to place. In both classic and contemporary urban sociology, the hypothesis generally is that urban life differs distinctly from rural, with city spaces—including marketplaces—being more fluid, less fixed and with weaker social ties compared to more rural spaces. However, the authors' comparison of a London market and a rural one in the south of England shows that such preconceptions are misconceiving. Instead, rural traders appear less attached to the market they were working in and had a weaker sense of community, mostly due to their different mobility patterns.

Markets are also often depicted in dominant public imaginations as 'local' sites. For example, the 'Love Your Local Market' campaign, which is annually organised in 24 different countries, serves to promote the idea of the proximity of the markets in terms of location, personal contact with traders and with products (NABMA, 2021). The last two contributions put this locality of the market in perspective. Joanna Menet and Janine Dahinden illustrate in Chapter 12 how traders at Swiss food markets consciously produce the image that both their products and themselves are from close proximity. Even if this is not the case, when clearly non-Swiss products are sold, these are represented as authentic local products from elsewhere. Hence, they conclude that markets are local only insofar as traders apply many strategies to bind their clients in proximity – but their locality is also conditioned by their embeddedness in a global identity politics where mobility, transnationalism and nationalism play a crucial role in the success of markets.

Emil van Eck, Rianne van Melik and Joris Schapendonk discuss in Chapter 13 how a European Union law from 2006 (the so-called Service Directive) has recently caused much concern among Dutch market traders. Their trading contracts can suddenly no longer be issued for an unlimited period but will become restricted in duration. Though this supposedly makes it easier for newcomers to enter the marketplace, the shorter contracts will also make it more difficult for traders to obtain bank credits and recoup their investment costs. The chapter shows that although most Dutch markets fall under the authority of municipal governments—and are hence 'local' affairs—the governance of marketplaces takes place on different spatial scales.

Routledge has a tradition of publishing titles on public space in general (Hou, 2010; Madanipour et al., 2014; Mehta & Palazzo, 2020) and (street) markets in particular (Evers & Seale, 2015; Seale, 2016; González, 2018). To pay tribute to these predecessors and indicate how the present edited volume builds upon this knowledge base, we invited the author of the latest Routledge book on marketplaces—Sara González—to write an afterword. In Chapter 14, she reflects upon the added value of using a mobility approach when studying marketplaces. The book closes with her call for an action research agenda that centres around marketplaces as essential urban spaces to foster more socially and environmentally just cities.

Development of this edited volume during COVID-19

This edited volume came into being after a broad call for chapters was launched in September 2019. Throughout various listservs and on social media, we invited researchers to submit contributions dealing with everyday marketplaces as dynamic and open entities with their institutional networks and constant flows of people, goods and ideas. Researchers were asked to link their contributions to one of the three themes (movements, representations and practices). After a careful selection, we welcomed a wide range of chapters—discussing different types of markets, spatial contexts and methodologies. Authors come from all over the world and discuss marketplaces in both the Global North and South, among others in, Brazil, Peru, Vietnam, South Africa, United Kingdom, Bulgaria, the Netherlands, India and Turkey.

In addition, the book contains contributions from many different disciplines, including the work of geographers, urban designers, sociologists, spatial planners and architects in various phases of their careers. As such, we feel the book generates multidisciplinary dialogues on markets, with the mobility perspective as a binding framework. Although it is a book on marketplaces, the contributors touch upon many themes relevant to other urban (public) spaces, such as urban transformation, ethnic diversity, social justice, place-making, translocality, institutional regulation and neoliberal urbanism.

It is important to note that full chapters were written in 2020 and 2021 amidst a global pandemic caused by the COVID-19 coronavirus. Though the virus is imperceptible, its impact has been very visible, perhaps most so in urban public spaces. In countries with very strict lockdowns, this resulted in empty streets and marketplaces, and spatial and temporal restrictions limiting the outdoor visits' frequency, duration and reach (Van Melik et al., 2021). In many regions of the world, marketplace-related interventions were ambiguous: while the dense, open and public character of markets were considered a major risk for contamination, they were also revalued for their significant role in the local food supply. Consequently, while some markets closed down completely, others were allowed to remain open under strict regulations and for only a limited number of (food) traders and customers (Van Eck et al., 2020).

Not only did the pandemic affect the object of investigation (i.e. marketplaces), but it also impacted our contributors' research activities. Some of the original fieldwork plans had to be adjusted; for example, Chapter 9 was initially planned to be a comparative study of marketplaces in Brazil and the United Kingdom, but this was impossible due to international travel restrictions. Fieldwork for Chapter 7 was not only impacted by COVID-19 but also limited by the devasting harbour explosion in Beirut that occurred in the same year. But also under less challenging spatial circumstances, all authors had to transition to homework and adjust to online teaching and research methods. We appreciate how everyone contributing to this edited volume has shown resilience and patience throughout the writing and editing process.

Notwithstanding these difficult circumstances, we do not feel this is primarily a book about markets in times of COVID-19, although some chapters do explicitly refer to the pandemic's effects (e.g. Chapters 6 and 10). First, one of our criteria was that fieldwork was more or less finished when we selected contributions at the beginning of 2020. The themes and methodologies of most chapters were therefore already well established before the pandemic emerged. Second, the pandemic certainly affected marketplaces, but in a way revealed nothing new. Many of the challenges discussed in this book, such as patterns of contestation and competition, were exacerbated or amplified rather than caused by the pandemic (Van Melik et al., 2021). As such, the pandemic merely revealed pre-existing narratives, tensions and pressures in the marketplace.

References

Abwe, F.G. (2020), Local Public Markets: The Empirical Evidence on Their Quantity and Quality in Arusha, Tanzania, https://healthbridge.ca/dist/library/FINAL_REPORT_ Local_Public_Markets_in_Arusha__Tanzania-2020.09.08.pdf (accessed November 25, 2021).

Adey, P. (2006), If mobility is everything then it is nothing: Towards a relational politics of (im)mobilities. *Mobilities*, 1(1), 75–94.

Adey, P. (2017), *Mobility*. London: Routledge.

Agnew, J.C. (1986), *Worlds Apart: The Market and the Theater in Anglo-American Thought, 1550–1750*. Cambridge: Cambridge University Press.

Albinsson P.A. & Perera, B.Y. (2012), Alternative marketplaces in the 21st century: Building community through sharing events. *Journal of Consumer Behaviour*, 11(4), 303–315.

Asante, L.A. & Helbrecht, I. (2020), Conceptualising marketplaces in Anglophone West Africa: A sexpartite framework. *GeoJournal*, 85, 221–236.

Breines, M.R., Menet, J. & Schapendonk, J. (2021), Disentangling following: Implications and practicalities of mobile methods. *Mobilities*, 16(6), 921–934. doi. 10.1080/17450101.2021.1942172

Chukwuemeka, V., Gantois, G., Scheerlinck, K., Schoonjans, Y. & Onyegiri, K. (2020), Embodying local identity as heritage in transition: The case study of Onistsha markets, Nigeria. In: Fouseki, K., Guttormsen, T.S. & Swensen, G. (Eds.), *Heritage and Sustainable Urban Transformations, Deep Cities* (pp. 55–66). New York: Routledge.

Cresswell, T. (2010), Towards a politics of mobility. *Environment and Planning D: Society and Space*, 28, 17–31.

Cresswell, T. (2011), Mobilities I: Catching up. *Progress in Human Geography*, 35(4), 550–558.

De La Pradella, M. (2006), *Market Day in Provence*. Chicago: The University of Chicago Press.

Endres, K.W. & Leshkowich, A.M. (2018), *Traders in Motion: Identities and Contestations in the Vietnamese Marketplace*. Ithaca: Cornell University Press.

Etzold, B. (2014), Migration, informal labour and (trans)local productions of urban space: The case of Dhaka's street food vendors. *Population, Space and Place*, 22(2), 170–184.

Evers, C. & Seale, K. (2015), *Informal Urban Street Markets: International Perspectives*. London: Routledge.

González, S. (2018), *Contested Markets, Contested Cities: Gentrification and Urban Justice in Retail Spaces*. London: Routledge.

Hall, P. (1988), *Cities of Tomorrow*. Oxford and New York: Basil Blackwell.

Hiebert, D., Rath, J. & Vertovec, S. (2015), Urban markets and diversity: Towards a research agenda. *Ethnic and Racial Studies*, 38(1), 5–21.

Hou, J. (2010), *Insurgent Public Space: Guerrilla Urbanism and the Remaking of Contemporary Cities*. New York: Routledge.

Jacobs, J. (1969), *The Economy of Cities*. New York: Vintage Books.

Janssens, F. & Sezer, C. (2013a), Marketplaces as an urban development strategy. *Built Environment*, 39(2), 168–311.

Janssens, F. & Sezer, C. (2013b), Flying markets: Activating public spaces in Amsterdam. *Built Environment*, 39(2), 245–260.

Jensen, O.B. (2009), Flows of meaning, cultures of movements: Urban mobility as meaningful everyday practice. *Mobilities*, 4(1), 139–158.

Kelley, V. (2016), The streets for the people: London's street markets 1850–1939. *Urban History*, 43(3), 392–411.

Kostof, S. (1999), *The City Shaped: Urban Patterns and Meaning Through History*. London: Thames and Hudson.

Lin Pang, C. & Sterling, S. (2013), From fake market to a strong brand? The Silk Street Market in Beijing. *Built Environment*, 39(2), 203–223.

Madanipour, A., Knierbein, S. & Degros, A. (2014), *Public Space and the Challenges of Urban Transformation in Europe*. New York: Routledge.

Marovelli, B. (2014), 'Meat smells like corpses': Sensory perceptions in a Sicilian urban marketplace. *Urbanities*, 4(2), 21–38.

Martin, D. (2003), 'Place-framing' as place-making: Constituting a neighborhood for organizing and activism. *Annals of the Association of American Geographers*, 93, 730–750.

Massey, D. (2001), Geography on the agenda. *Progress in Human Geography*, 25(1), 5–17.

Massey, D. (2005), *For Space*. London: Sage.

Mayhew, H. (1985), *London Labour and the London Poor*. London: Penguin Books.

Mehta V. & Palazzo D. (2020), *The Routledge Companion to Public Space*. London: Routledge.

Michelon, B. (2009), The Local Market in Kigali as Controlled Public Space: Adaptation and Resistance by Local People to Modern City Life. PhD Seminar on Public Space, Delft, the Netherlands. http://resolver.tudelft.nl/uuid:e7d458d2-18d1-459b-a666-963e27c4e713 (accessed November 25, 2021).

Morales, A. (2009), Public markets as community development tools. *Journal of Planning Education and Research*, 28(4), 426–440.

Morales, A. (2011), Marketplaces: Prospects for social, economic, and political development. *Journal of Planning Literature*, 26(1), 3–17.

NABMA (2021), Love Your Local Market 2021, https://nabma.com/love-your-local-market-2021/ (accessed August 30, 2021).

Neethi, P., Kamath, A. & Paul, A.M. (2021), Everyday place making through social capital among street vendors at Manek Chowk, Gujarat, India. *Space and Culture*, 24(4), 570–584.

Nikšič, M. & Sezer, C. (2017), Public space and urban culture. *Built Environment*, 43(2), 165–172.

Polyák, L. (2014), Exchange in the street: Rethinking open-air markets in Budapest. In: Madanipour, A., Knierbein, S. & Degros, A. (Eds.), *Public Space and the Challenges of Urban Transformation in Europe* (pp. 48–59). New York: Routledge.

Pottie-Sherman, Y. (2013), Vancouver's Chinatown night market: Gentrification and the perception of Chinatown as a form of revitalisation. *Built Environment*, 39(2), 172–189.

Qiang, S. (2013), Market hierarchies produced by scale-structure: Food markets in the third ring of Beijing. *Built Environment*, 39(2), 297–311.

Schappo, P. & Van Melik, R. (2017), Meeting at the marketplace: On the integrative potential of The Hague Market. *Journal of Urbanism*, 10(3), 318–332.

Seale, K. (2016), *Markets, Places, Cities*. New York/London: Routledge.

Seligman, L.J. & Guevara, D. (2013), Occupying the center: Handicraft vendors, cultural vitality, commodification, and tourism in Cusco, Peru. *Built Environment*, 39(2), 203–223.

Sennett, R. (1998), *The Spaces of Democracy: 1998 Raoul Wallenberg Lecture*. Michigan: University of Michigan.

Sezer, C. (2020), *Visibility, Democratic Public space, and Socially Inclusive Cities: The Presence and Changes of Turkish Amenities in Amsterdam*. Delft: A+BE Architecture and the Built Environment.

Sheller, M., & Urry, J. (2006), The new mobilities paradigm. *Environment and Planning A: Economy and Space*, 38, 207–226.

Stobart, J. & Van Damme, I. (2016), Introduction: Markets in modernization. Transformations in urban market space and practice, c.1800–c.1970. *Urban History*, 43(3), 358–371.

Ünlü-Yücesoy, E. (2013), Constructing the marketplace: A socio-spatial analysis of past marketplaces of Istanbul. *Built Environment*, 39(2), 190–202.

Urbact Markets (2015), *Urban Markets: Heart, Soul and Motor of Cities*. Barcelona: City of Barcelona.

Urry, J. (2007), *Mobilities*. Cambridge: Polity Press.

Van Eck, E. (2021), "That market has no quality": Performative place frames, racialisation, and affective re-inscriptions in an outdoor retail market in Amsterdam. *Transactions of the Institute of British Geographers*. doi. 10.1111/tran.12515.

Van Eck, E., Van Melik, R. & Schapendonk, J. (2020), Marketplaces as public spaces in times of the COVID-19 coronavirus outbreak: First reflections. *Tijdschrift voor Economische en Sociale Geografie*, 111(3), 373–386.

Van Melik, R., Filion, P. & Doucet, B. (2021), *Global Reflections on COVID-19 and Urban Inequalities: Public Space and Mobility (Volume 3)*. Bristol: Bristol University Press.

Van Melik, R. & Spierings, B. (2020), Researching public space: From place-based to process-oriented investigations. In: Mehta V. & Palazzo D. (Eds.), *The Routledge Companion to Public Space* (pp. 16–27). London: Routledge.

Watson, S. (2009), The magic of the marketplace: Sociality in a neglected public space. *Urban Studies*, 46(8), 1577–1591.

2 Hanoi's street vendors on the move

Itinerant vending tactics and mobile methods in the Socialist Republic of Vietnam

Celia Zuberec and Sarah Turner

Introduction

In 2010, Hanoi, the political capital of the Socialist Republic of Vietnam, celebrated its official 1,000th anniversary, considered by many as a momentous occasion recognising the dynamic history of the city (Drummond, 2012). This milestone was also seen as an important reason to highlight the city's rapid modernisation and its trajectory towards becoming a truly global city (Karis, 2017). Prior to this anniversary, Hanoi's municipal government had launched several campaigns and policies to guide the city along this path of "'urban modernization' (*hiện đại hóa*) and 'urban civilization' (*văn minh đô thị*)" (Gibert & Segard, 2015: 7). Most drastically, in 2008, the prime minister expanded the capital's limits, tripling the city's surface area (Leducq & Scarwell, 2018). That same year, Hanoi's municipal government announced a ban on street vending in 63 main streets and 48 public spaces, with street vendors, the individuals at the core of this chapter, deemed "backward, inefficient and detrimental to national development schemes" (Cross, 2000: 40). Not surprisingly, this ban has had significant negative implications on the ability of thousands of vendors in Hanoi to secure viable livelihoods (Turner & Schoenberger, 2012).

Yet the street vending ban has not been experienced uniformly by the city's vendors, with its implementation disproportionately affecting itinerant vendors, most commonly rural-to-urban migrants from peri-urban areas (Jensen & Peppard, 2003). In comparison, fixed-stall vendors, largely long-term Hanoi residents, are often able to bypass the repercussions of the ban by leveraging their social and financial capital. These fixed-stall vendors frequently pay informal 'fees' to ward officials, with whom they often have long-established social relations, to maintain access to the city's streets (Koh, 2006; Turner & Schoenberger, 2012). Without access to such social networks, operating a fixed stall is generally outside of the means of migrant vendors. These individuals not only face fines and harassment from the city's ward officials but are also commonly marginalised by long-term Hanoi residents who regard them as 'backwards' and 'unmodern' (Turner & Schoenberger, 2012; Eidse et al., 2016; own fieldwork).

In this chapter, we aim to analyse the mobility tactics that itinerant street vendors in Hanoi draw upon to access potential customers and evade prosecution. While doing so, we highlight how vendors' knowledge of the city and

DOI: 10.4324/9781003197058-2

access to customers is "constructed in and through mobile interactivity" (Brown & Durrheim, 2009: 916). We outline our conceptual framework next, drawing on mobility debates that are centrally concerned with how "place and context" shape experiences of movement (Curl et al., 2018: 174). We then emphasise how mobile methods can uncover intricate relationships among people, perceptions, experiences and specific places. We detail the particular 'go-along' method we adopted, that complemented our stationary conversational interviews and participant observation. This 'go-along' approach provided us with nuanced understandings of vendor tactics and everyday decision-making and allowed us to analyse the spatiality, temporality and significance of their routines. While finding clear evidence that "speeds, slownesses, and immobilities are all related in ways that are thoroughly infused with power and its distribution" (Cresswell, 2010: 21), we also explore how vendors draw on a range of dynamic everyday tactics (following De Certeau, 1984), to continue to reach customers and avoid state sanctions.

Critical mobilities scholarship and the 'mundane'

As an interdisciplinary subject of study, "the 'new mobilities' paradigm" (Sheller & Urry, 2006: 208) aims to specifically explain and analyse the movement of "people, objects, information and ideas" (Urry, 2007: 43). While transportation studies, feminist geographies and migration scholarship have all taken movement, displacement and mobility into consideration for decades, the mobility turn is argued to mark the beginning of more comprehensive attempts to understand "the entanglement of movement, representation and practice" (Cresswell, 2010: 19; see also Cresswell, 2006). Important to these attempts are understandings of how mobility and immobility are differentially accessed, experienced and embodied (Binnie et al., 2007). Characteristics such as "race, class, gender, sexuality, disability, nationality [and] age" must be acknowledged in order to understand the freedoms and restrictions placed on different bodies (Oswin, 2014: 85). Furthermore, critical mobilities scholars argue that attention must be paid to a wide range of forms and scales of movement (Binnie et al., 2007, Cresswell, 2011). For instance, to complement the increasing studies of '21st-century forms of movement', such as driving, air travel and virtual mobility (Cresswell, 2010), "banal or mundane mobilities" should be given proportionally more attention (Binnie et al., 2007: 165). These banal mobilities—such as walking, pushing a hand cart or biking—are important for understanding power relations and how meaning is ascribed to place (Binnie et al., 2007).

Cresswell (2010: 17) has advanced a framework for understanding "the politics of mobility" that is sensitive to social relations, scale and form. His approach identifies six elements that contribute to differential experiences of mobility—namely, "motive force, velocity, rhythm, route, experience and friction". He suggests that each contributes to the production of "mobile hierarchies and the politics of mobility" and that through their analysis one might start to determine how "mobility becomes political" (Cresswell, 2010: 22). We draw upon these constituent parts of mobility here, especially focusing on rhythms, routes, experiences and frictions, as we analyse the practices of Hanoi's migrant street vendors as they negotiate access to Hanoi's public spaces for their livelihoods.

Mobile methods

The emergence of the new mobilities paradigm and corresponding rise of mobilities scholarship has created a growing interest in mobile methods (Sheller & Urry, 2006). Mobile methods provide insights into the ways "place and context" shape experiences of movement (Curl et al., 2018: 174) and how "knowledge is constructed in and through mobile interactivity" (Brown & Durrheim, 2009: 916). Often designed as a combination of participant observation and interviewing, these methods allow researchers to gain access to participants' experiences and insights while simultaneously observing their practices and interactions *in situ* and on the move (Kusenbach, 2003; Carpiano, 2009; Block et al., 2019). Mobile methods involve accompanying a research participant through a "physical and social environment" (Battista & Manaugh, 2017: 12) and might include walking (e.g. Anderson, 2004; Battista & Manaugh, 2017; Ireland et al., 2019; Glenn et al., 2020; Kidman et al., 2020), running (Cook et al., 2016), cycling (Curl et al., 2018, Xie & Spinney, 2018), taking public transportation (Wegerif, 2019), driving a motorised vehicle (Brown & Durrheim, 2009) or riding a motorbike (Turner, 2020). Such scholars argue that by immersing themselves in the practices of their participants in these ways, they are able to better access and analyse participant experiences of (im)mobility (Block et al., 2019; Ireland et al., 2019; Wegerif, 2019; Burns et al., 2020).

Far from being uniformly experienced, mobility is "stratified and conditioned by access to resources and by one's identity" (McCann, 2011: 121). Mobile methods have therefore been used to capture how children (Block et al., 2019), young adults (Glenn et al., 2020, Kidman et al., 2020) and the elderly (Curl et al., 2018) experience (im)mobility. They have also provided a helpful tool for accessing the contextually sensitive perceptions of ethnic minorities (Brown & Durrheim, 2009; Carpiano, 2009), indigenous communities (Kidman et al., 2020), women (Xie & Spinney, 2018), informal workers (Turner, 2020), people experiencing homelessness (Martini, 2020) and individuals facing health concerns (Burns et al., 2020; Ireland et al., 2019). What soon becomes clear is that mobile methods have contributed greatly to discussions of the ways in which mobility occurs within complex dynamics of "politics, power, and ideology" (Cresswell, 2006: 55), a proposition that we wanted to explore further with regards to itinerant street vendors in Hanoi.

Research approach: Route maps and go-along interviews

During the summer of 2018, first author Celia completed 35 stationary, semi-structured interviews and participant observation with itinerant street vendors in Hanoi. To complement these findings, an additional four day-long, go-along interviews were undertaken to gain a nuanced understanding of the everyday dynamics of route choices and ward official negotiations and how vendors attach meaning to specific places and routes. The go-along interviews took the form of either 'walk-alongs', with Celia walking alongside a participant, or 'ride-alongs' whereby she drove a motor-scooter alongside the street vendor as the vendor cycled or motor-biked between trading locations. Adopting a go-along interview approach allowed us to simultaneously ask questions and observe the vendors as they carried out their daily routines.

The go-along interviews started when Celia met the vendors between 3:30 and 5:00 am at either Quảng Bá or Long Biên markets, depending on the products they were purchasing to resell that day. She then accompanied the vendors on their usual routes with no explicit instructions provided in an attempt to minimise the influence of her presence, an approach sometimes called a 'natural' go-along (Kusenbach, 2003). Conversational interviews were undertaken along the route when the vendor was not engaged in actively stocking their goods or selling. These included pointing to specific features of the environment to prompt further discussion (cf. Carpiano, 2009). With the vendor's permission, we also mapped their route through the city using the smartphone application Strava. This allowed us to contextualise the interview data collected, increasing our understandings of the influence of place on the participants, experiences and allowing us to create narrative maps such as Figure 2.2.

We also asked all of the 35 vendors with whom semi-structured interviews were completed precise questions detailing the routes they tended to follow on a daily basis. We then plotted these routes on base maps using QGIS. The route maps of three vendors are portrayed in Figure 2.1, and take place in Hanoi's 'Old

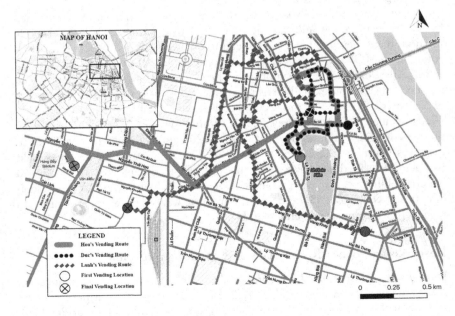

Figure 2.1 Three itinerant vendors' daily routes.

Source: authors.

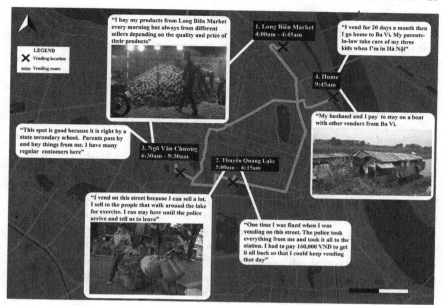

Figure 2.2 Female vendor Vinh's narrative map.
Source: authors.

Quarter', the city's historical centre and now its main tourism area. The start point of each route indicates the first location the vendor goes to sell their products after purchasing them at a wholesale market, while the end point denotes the final place they vend before returning home for the day. These data are further supported by over 75 interviews with vendors in Hanoi since 2010 by second author Sarah.

Itinerant street vendor mobility rhythms, routes, experiences and frictions

Drawing on Cresswell's (2010) core components of mobility—especially rhythms, routes, experiences and frictions—we turn here to examine the practices of Hanoi's migrant street vendors as they navigate access to Hanoi's public spaces to maintain their livelihoods.

Daily rhythms and routes

Figure 2.1 provides a visual sample of the ways participants designed their daily rhythms and routes to balance their needs to meet customers and secure their livelihoods on the one hand, with official attempts to channel their motion into "producing correct mobilities" on the other (Cresswell, 2010: 24). These routes are thus the product of both "choice" and "constraint" (Hanson, 2010: 14). We should note that route length does not necessarily correspond to time spent vending, as some vendors chose to stay longer in specific places to trade, as well as to avoid officials. Importantly, more than 75% of participants suggested that they follow the same route every day in order to develop relationships with regular customers. That said, many vendors will make temporary modifications to their routes in response to changes in weather (for example heavy rain or high temperatures), or, less frequently, permanent changes to account for the increased presence of ward police in specific areas.

Figure 2.1 illustrates the daily routines of three vendors. Duc's route is the shortest by length of time (all vendor names are gender-appropriate pseudonyms). Duc sells fruit that he purchases each morning at a large wholesale market, Long Biên. He then weaves his way through the Old Quarter between 4:00 and 10:00 am trading surreptitiously on a number of streets covered by the 2008 vending ban. Meanwhile, Lanh prepares sticky rice each morning, which she then supplies to a number of regular customers along a consistent route, as well as chance encounters with other purchasers during her nine hours of daily trade. Hoa buys flowers from Quảng Bá Flower Market early each morning, which she then sells for ten hours each day, walking very slowly and stopping often with her pushbike. Together, these routes outlined in Figure 2.1 (and others not presented here) illustrate a diversity of street networks and other public spaces itinerant vendors access for their trade.

The vendors' routes are also shaped by specific temporalities and over two-thirds of the 35 vendor interviewees match their vending routes to the schedules of their customers. For some, this involves arriving at a recreation area, such as a lakeside park, early in the day to catch residents ending their morning exercises or standing alongside busy roads during rush hour. Vendors also take advantage of a lull in customers around midday to rest, eat lunch or move to their next location. These vendors are highly attuned to the rhythms of Hanoi's residents and do what they can to exploit these rhythms for their economic gain.

The day of the week and time of the month also affect some vendors' routes. For instance, interviewee Nhung explained that she has two separate routes: one for Monday to Friday, and one for the weekend when the 'walking streets' are open (specific streets in the Old Quarter blocked to traffic on the weekends to entice both locals and tourists to the area). Further, nearly half the vendors marked a difference in their vending practices on the 1st and 15th of the lunar calendars. On these days, many city residents buy fruit and flowers to offer to their ancestors, hence creating higher demand for some vendors' goods. Leading up to these days vendors selling these products start work earlier to ensure they can purchase the best merchandise from wholesale markets and maximise their daily sales. Overall, the average workday of our interviewees was ten hours, with some working as long as 16 hours and none working less than five.

Drawing on the specific data collected during the four go-along interviews, we also created narrative maps to represent more nuanced understandings of the lived experiences of vendors in their socio-spatial environments. The narrative depicted here is that of Vinh, an itinerant produce vendor (Figure 2.2). In her walking-while-talking interview, Vihn explained that she tries to follow the same route every day to ensure that she meets her regular customers. However, she will readily modify her route if she sees ward officials approaching who might confiscate her fruit. Her map shows that the temporality of her vending route is also shaped by constraints and opportunities. She needs to leave a lakeside area before ward officials arrive at 6:30 am to avoid fines, and she needs to be at an elementary school in time to sell to children's parents arriving or leaving. The narrative map also situates these daily mobilities within larger temporal and spatial scales. For example, Vinh's map shows her circular migration to and from the countryside. She explained, "I vend for 20 days a month and then I go home to Ba Vì" where she attends to chores on her family farm.

Mobility frictions

By attempting to "enhance the mobilities" of 'modern' cars and motorbikes, the 2008 vending ban has "[reinforced] the immobilities" of vendors (Block et al., 2019: 1296). As a result, none of the 35 vendors felt they had easy access to Hanoi's public spaces for their work. As Turner and Schoenberger (2012) have noted, the street vending ban is enforced at the urban administrative unit level of the ward and its implementation is often at the discretion of ward officials. Interviewees noted that they avoid specific areas where ward officials are known to hand out particularly high fines, wards with a larger presence of officials and places where they have previously been caught. Truc, a woman vendor selling flowers, explained "I avoid Thành Công area because I was fined 150,000 [USD$6.50] twice in five days there. There are so many officials there and they work very hard". Similarly, an increased presence of ward officials in the Old Quarter has caused some vendors to make permanent alterations to their routes. Lanh, a 30-year-old sticky rice vendor explained, "I used to vend more in the Old Quarter where I could sell my products for a higher price. Now there are too many officials so I vend here [itinerantly around Ngô Sĩ Liên outdoor market; 2km from Old Quarter] and I have to sell my product for a lower price".

The 2008 street vending ban and subsequent anti-vendor campaigns have encoded the vendors' rhythms and speeds as inefficient and unmodern (Cresswell, 2010). Due to these narratives, vendors' slow movement is seen as undesirable by ward officials, and the list of negative encounters vendors have endured at the hands of officials was extensive. Many recounted officials routinely confiscating their baskets, scales and produce, keeping their identity cards or driver's licences, insulting and yelling at them, holding their motorbikes for a bribe and fining them. Hau, a woman selling chilies, limes and taro, shared some of her worst confrontations with ward officials:

> One time when I was caught they took away all my products until 6:30 pm. Another time in a different ward when I was caught I was so scared that I peed myself. They were going to take away all of my products again!

Hau's fears are testament to the fact that the fees to have goods released can cripple a vendor's livelihood, with fines often in the same range as average daily profits, at VND150,000–VND200,000 (USD$6.50–8.50).

Vendors also highlighted how the severity of bribes or fines depends on the character of the specific official. Binh, a male pineapple vendor from Hưng Yên province explained:

> Sometimes when the ward police stop me they fine me and give me a ticket. Other times I'm fined, but not given a ticket so the police can "eat my money" [keep the funds]. Many of the ward police have been to jail; they are bad and mean.

There was a general consensus among vendors that the ban's implementation had become more strict every year, with a surge in severity since the initiation of a 'Clean up the Sidewalk Campaign' in 2017 targeting a range of activities on the city's sidewalks. Hanson (2010) argues that the meaning attached to movement can only be understood by considering it within its socio-political context. Here it is clear that state discourses relating to the movement of vendors have impacted their ability to move freely through the city (see also Blomley, 2011). Lanh, a 30-year-old woman vendor explained, "When I first started vending, if I stopped to rest on the sidewalk and wasn't selling to anyone I wouldn't be fined. Now if I stop on the sidewalk I'll be fined even if I'm not selling!"

Interviewees added that restrictions tend to be more forcefully upheld on national holidays, during important parliamentary meetings or when an overseas head of state visits. Many also noted that fines are more common at the start of the working week. Thi, a young fruit vendor from Ba Vì detailed, "On Mondays and Tuesdays the police start earlier and work harder to catch us, but on the weekends there's less police". Two-thirds of interviewees also noted a correlation between the severity of the ban's enactment and lunar holidays. They felt the ban was implemented *less* rigorously on the 1st and the 15th of the lunar calendar. Thao explained, "The police work less hard because they understand people must buy fruit to offer their ancestors", while Bich noted that even though "on the 1st and the 15th the police still want to get rid of me, if they catch me they may forgive me". This adds an unexpected cultural dimension to the temporalities of how mobility frictions play out across the city.

Public/private transcripts and the politics of mobility

Political scientist James C. Scott (1990) has proposed two useful terms for investigating the political struggles of subordinate groups that we draw on here. First, the *public transcript* refers to the "open interaction between subordinates and those who dominate" (Scott, 1990: 2). It describes the way that subordinated individuals or groups modify their behaviours and discourses to align with the "hegemony of dominant values" when interacting with elites or those in positions of power (ibid: 4). Second, the *hidden transcript* consists of "those offstage speeches, gestures, and practices that confirm, contradict, or inflect what appears in the public transcript" (Scott, 1990: 4–5). In line with Scott's conceptual ideas, we found there to be important ways by which itinerant vendors engage in forms of everyday compliance, engaging in acts that may reinforce the system of authority, but doing so without the desire to sustain its domination (Kerkvliet, 2009). As an example, some interviewees explained that they give money to ward officials either monthly or on national holidays not as a fine or a fee but a 'gift'. Dũng elaborated:

> When there is a national event, the other vendors and I contribute money to the ward police. This is a customary practice; us vendors work in a ward where we're taken care of by the police. Even if police ask the vendors to move, they still take care of the ward. You want to have a good relationship with the police.

Although this deed—or public transcript—could be interpreted as *support* for ward officials (Kerkvliet, 2009; Scott, 1990), the discrepancy between this act and the critical words that this vendor then shared about the unfair treatment by ward officials reveals that it is more a matter of compliance and that a hidden transcript also exists (Kerkvliet, 2009). Indeed, "the safest and most public form of political discourse is that which takes its basis in the flattering self-image of elites" (Scott, 1990: 18).

Itinerant street vendors noted a number of ways that they navigate, negotiate and/or resist the terms of their subordination (Scott, 1990; Kerkvliet, 2009; Johansson & Vinthagen, 2016). Building on the work of Turner and Schoenberger (2012) and Eidse et al. (2016), we argue that vendors do not simply acquiesce to the terms of the street vending ban. Nearly every vendor described how they have to be on the lookout for ward officials and be ready to move quickly should they see them. Many explained the importance of putting their weighing scales back on their bike after each sale, to avoid leaving them behind if they have to flee. As vendor Tuyen explained, "In order to avoid the police I need to always walk around and never put my carrying pole or basket down. I need to be able to leave right away".

The interviewees' tactics of everyday resistance also have a specific "temporalisation" (Johansson & Vinthagen, 2016: 427), with some explaining that they have learned to coordinate their vending routes with ward official schedules. For instance, at an inner-city lakeside where itinerant venders cluster to sell in the early morning, they know to clean up and leave before the police arrive around 6:30 am, but the same vendors sometimes then return to the site after ward officials have moved on to other areas. While this presents a public transcript of conformity, the hidden transcript is characterised by defiance (Scott, 1990). This "false compliance" (Scott, 1989: 9) happens repeatedly throughout the city over the course of each day. For instance, female vendor Truc explained, "I know the schedule of the police so I can avoid being caught". Similarly, other vendors recounted that they strategically determine when to enter wards with expensive fines, targeting late nights or weekends, when there are fewer ward officials present.

Finally, it is also clear that the social networks vendors have developed among themselves contribute in important ways to their ability to simultaneously resist the police and secure their livelihoods. Many vendors explained that they either yell to nearby vendors or use their cell phones to warn vendor friends and family members that ward officials are in the area. As such, street vendor interviewees draw on numerous tactics to avoid fines and retain the mobility that underscores their livelihoods.

Conclusion

While drawing on mobile methods, semi-structured interviews, observations and narrative mapping, we have highlighted a range of nuances involved in the spatial and temporal tactics that Hanoi's itinerant vendors engage with to create resilient livelihoods in the face of significant obstacles. In this way, our findings confirm past research (Turner & Schoenberger, 2012) that while long-term Hanoi residents can draw on social capital with ward officials and financial capital in order to secure access to trade from fixed stalls on the city's streets, migrant itinerant vendors must be more tactical and covert in their actions. Itinerant vendors utilise their in-depth knowledge of the rhythms and routines of numerous actors to maximise sales while minimising harassment from ward officials. Their routes are produced through a careful negotiation between their desire to meet and sell to regular customers and find new ones, and their need to change routines and be ready to flee should they encounter ward officials. Further, our research results indicate that when tactics of resistance and evasion fail, vendors often resort to politics of compliance, showing deference and remorse to ward police in order to limit penalties.

Reflecting on the mobile methods that we utilised here, we found that mobile go-along interviews were a particularly valuable approach for revealing patterns, causes and impacts of the mobilities and immobilities of Hanoi's itinerant vendors. Indeed, like other scholars, we found that go-along interviews allowed us to produce richer data than is possible with traditional sedentary methods (Evans & Jones, 2011; Burns et al., 2020). In large part, this was due to the situatedness of the go-along interviews that placed us alongside our participants within the location of interest. We were therefore able to watch in real time as vendors responded to various elements within the landscape and how, at times, that translated into modifications to their trade practices. Significantly, through these go-along interviews, the temporality and spatiality of the vendors' mobility became more obvious, allowing us to ask place-specific questions and reinforcing Jones et al.'s (2008: 2) argument that there is a relationship between "what people say" and "where they say it".

We also greatly appreciated how the go-alongs allowed us to shift some control over the interview process to the vendors (Block et al., 2019; Carpiano, 2009; Burns et al., 2020). Our experiences in this regard reflected those of Brown and Durrheim (2009: 917), who found that walking alongside their research participants positioned them as a "(co)producer of dialogue", and that of Wegerif (2019: 6), who found this process helped "build solidarity and a comfortable rapport". As we accompanied the vendors along their daily routes, they were in many ways directing the interview. Not only did this provide us with more accurate, contextually specific data but also allowed us to modify the power dynamics between us.

Overall, it is clear that the obstacles vendors face are frequently linked to the politicisation of vendor mobility. As the Vietnamese state moves forward with its global city vision and modernisation campaigns, vendors' slow, stop-start mobility is increasingly positioned as incompatible. It is therefore imperative that we continue to gain a better understanding of "power relations embedded in various forms of mobility/immobility in various social and geographic contexts" (Hanson, 2010: 11). By doing so, we can make sure that dominant renditions of mobilities—those propagated by the Vietnamese state—are not the only narratives being recorded and that alternative knowledges of mobility, such as the 'banal' everyday mobilities of itinerant vendors, are highlighted, better recognised and perhaps one day even supported to improve mobility justice and access to the city's streets.

References

Anderson, J. (2004), Talking whilst walking: A geographical archaeology of knowledge. *Area*, 36(3), 254–261.

Battista, G.A. & Manaugh, K. (2017), Using embodied videos of walking interviews in walkability assessment. *Transportation Research Records*, 2661(1), 12–18.

Binnie, J., Edensor, T., Holloway, J., Millington, S. & Young, C. (2007), Mundane mobilities, banal travels. *Social & Cultural Geography*, 8(2), 165–174.

Block K., Gibbs L. & MacDougall, C. (2019), Participant-guided mobile methods. In: Liamputtong, P. (Ed.), *Handbook of Research Methods in Health Social Sciences* (pp. 1291–1302). Singapore: Springer.

Blomley, N. (2011), *Rights of Passage: Sidewalks and the Regulation of Public Flow*. Milton Park: Routledge.

Brown, L. & Durrheim, K. (2009), Different kinds of knowing: Generating qualitative data through mobile interviewing. *Qualitative Inquiry*, 15(5), 911–930.

Burns, R. Gallant, K.A., Fenton, L. White, C. & Hamilton-Hinch, B. (2020), The go-along interview: A valuable tool for leisure research, *Leisure Sciences*, 42(1), 51–68.

Carpiano, R.M. (2009), Come take a walk with me: The "go-along" interview as a novel method for studying the implications of place for health and well-being. *Health & Place*, 15, 263–272.

Cook, S. Shaw, J. & Simpson, P. (2016), Jography: Exploring meanings, experiences and spatialities of recreational road-running. *Mobilities*, 11(5), 744–769.

Cresswell, T. (2006), *On the Move*. New York: Routledge.

Cresswell, T. (2010), Towards a politics of mobility. *Environment and Planning D: Society and Space*, 28(1), 17–31.

Cresswell, T. (2011), Mobilities I: Catching up. *Progress in Human Geography*, 35(4), 550–558.

Cross, J.C. (2000), Street vendors, modernity and postmodernity: Conflict and compromise in the global economy, *International Journal of Sociology and Social Policy*, 1(2), 30–52.

Curl, A., Tilley, S. & Cauwenberg, J.V. (2018), Walking with older adults as a geographical method. In: A. Curl & C. Musselwhite (Eds.), *Geographies of Transport and Ageing* (pp. 171–195). Basingstoke: Palgrave Macmillan.

de Certeau, M. (1984), *The Practice of Everyday Life*. Berkeley: University of California Press.

Drummond, L.B.W. (2012), Middle class landscapes in a transforming city: Hanoi in the 21st century. In: V. Nguyen-Marshall, L. Drummond & D. Bélanger (Eds.), *The Reinvention of Distinction* (pp. 79–93). Dordrecht: Springer.

Eidse, N., Turner, S., & Oswin, N. (2016), Contesting street spaces in a socialist city: Itinerant vending-scapes and the everyday politics of mobility in Hanoi, Vietnam. *Annals of the American Association of Geographers*, 106(2), 340–349.

Evans, J. & Jones, P. (2011), The walking interview: Methodology, mobility and place. *Applied Geography*, 31, 849–858.

Gibert, M. & Segard, J. (2015), Urban planning in Vietnam: A vector for a negotiated authoritarianism? *Justice Spatiale – Spatial Justice*, 8, Retrieved From: http://www.jssj.org/article/lamenagement-urbain-au-vietnamvecteur-dun-autoritarisme-negocie/

Glenn, N.M., Frohlich, K.L. & Valléed, J. (2020), Socio-spatial inequalities in smoking among young adults: What a 'go-along' study says about local smoking practices. *Social Science & Medicine*, 253, 1–8.

Hanson, S. (2010), Gender and mobility: New approaches for informing sustainability. *Gender, Place & Culture*, 17(1), 5–23.

Ireland, A.V., Finnegan-John, J., Hubbard, G., Scanlon, K., & Kyle, R.G. (2019), Walking groups for women with breast cancer: Mobilising therapeutic assemblages of walk, talk and place. *Social Science & Medicine*, 231, 38–46.

Jensen, R. & Peppard, D. (2003), Hanoi's informal sector and the Vietnamese economy: A case study of Roving Street vendors. *Journal of Asian and African Studies*, 38(1), 71–84.

Johansson, A. & Vinthagen, S. (2016), Dimensions of everyday resistance: An analytical framework. *Critical Sociology*, 42(3), 417–435.

Jones, P., Bunce, G., Evans, J., Gibbs, H. & Hein, J.R. (2008), Exploring space and place with walking interviews. *Journal of Research Practice*, 4(2) 1–9.

Karis, T. (2017), The wrong side of the tracks. *City*, 21(5), 663–671.

Kerkvliet, B.J.T. (2009), Everyday politics in peasant societies (and ours). *The Journal of Peasant Studies*, 36(1), 227–243.

Kidman, J., MacDonald, L, Funaki, H., Ormond, A., Southon, P. & Tomlins-Jahnkne, H. (2020), 'Native time' in the white city: Indigenous youth temporalities in settler-colonial space. *Children's Geographies*, 19(1), 24–36.

Koh, D.W.H. (2006), *Wards of Hanoi*, Singapore: ISEAS-Yusof Ishak Institute.

Kusenbach, M. (2003), Street phenomenology: The go-along as ethnographic research tool. *Ethnography*, 4(3), 455–485.

Leducq, D. & Scarwell, H.J. (2018), The new Hanoi: Opportunities and challenges for future urban development. *Cities*, 72, 70–81.

Martini, N. (2020), Using GPS and GIS to enrich the walk-along method. *Field Methods*, 32(2), 180–192.

McCann, E. (2011), Urban policy mobilities and global circuits of knowledge: Toward a research agenda. *Annals of the Association of American Geographers*, 101(1), 107–130.

Oswin, N. (2014), Queer theory. In: P. Adey, D. Bissell, K. Hannam, P. Merriman & M. Sheller (Eds.), *The Routledge Handbook of Mobilities* (pp. 85–93). New York: Routledge.

Scott, J.C. (1989), Everyday forms of resistance. In: F.D. Colburn (Ed.), *Everyday Forms of Peasant Resistance* (pp. 3–33). London/New York: Routledge.

Scott, J.C. (1990), *Domination and the Arts of Resistance: Hidden Transcripts*. New Haven: Yale University Press.

Sheller, M. & Urry, J. (2006), The new mobilities paradigm. *Environment and Planning A*, 38, 207–226.

Turner, S. (2020), Informal motorbike taxi drivers and mobility injustice on Hanoi's streets: Negotiating the curve of a new narrative. *Journal of Transport Geography*, 85, 1–8.

Turner, S. & Schoenberger, L. (2012), Street vendor livelihoods and everyday political in Hanoi, Vietnam: The seeds of a diverse economy? *Urban Studies*, 49(5), 1027–1044.

Urry, J. (2007), *Mobilities*. Cambridge: Polity Press.

Wegerif, M.C.A. (2019), The ride-along: A journey in qualitative research. *Qualitative Research Journal*, doi 10.1108/QRJ-D-18-00038.

Xie, L. & Spinney, J. (2018), "I won't cycle on a route like this; I don't think I fully understood what isolation meant": A critical evaluation of the safety principles in Cycling Level of Service (CLoS) tools from a gender perspective. *Travel Behaviour and Society*, 13, 197–213.

3   Rhythmic encounters in an Indian marketplace

Kiran Keswani

Introduction

In his delightful essay "The Crowd", R. K. Narayan—an Indian writer who always highlighted the social context and everyday life of his characters—sums up the culture of a typical Indian street or a marketplace. He says, "Any crowd interests me: I always feel that it is a thing that deserves precedence over any other engagement. I always tell myself that an engagement can wait, but not the crowd" (1989, p. 37). While studying marketplace, one realises through observing and experiencing them that they are not just simple selling spaces but receptacles that hold objects, experiences and people in an unending, fascinating churn. On an everyday basis, there are flower vendor spaces inextricably woven with the vegetable vendor spaces; street corners become either the confluence or receding of the same vendor groups. On a periodic basis, there is the emergence of an entire street full of kite makers, usually a month before the harvest festival or the rising crescendo of firecracker sales as the festival of lights draws near. This chapter is about such a marketplace and has two parts. The first part describes the movement patterns in the marketplace, of delivery agents of vendors selling *chai* (tea) to formal shop owners in the neighbourhood. The second part looks at a mapping of the *everyday routes* of this group of actors at different times of the day.

In India, as elsewhere, people frequent marketplaces for their day-to-day needs, and they encounter others as they come to the market in the morning or evening. There are both business and social encounters that happen here. Every encounter is related to a ritual of one kind or another—either a ritual related to how WE move in relation to our daily needs, a ritual related to how OTHERS move in relation to us, or a ritual related to how we move in relation to how the EARTH moves around the sun, creating shadows which lend value to a place for social interaction, especially in a tropical country. There are multiple actors within the marketplace environment—those who come to buy, those who come to sell and those who are onlookers of the rituals that unfold one day after another. A marketplace can be a place where different urbanites interact with teashops, retail stores, barbershops and the like, becoming interesting sites to investigate social structures and social interactions (Berreman, 1972). But, more than anything else, these are places where people come to be together for some

DOI: 10.4324/9781003197058-3

part of the day. As Narayan (1989) says at the end of his essay, "[F]or human beings the greatest source of strength lies in each other's presence".

Urban design scholars have looked at public spaces such as streets and marketplaces in different ways. Janssens and Sezer (2013) suggest that marketplaces are representative of the everyday life of the city as they facilitate flows of people, goods and information. Similarly, in the context of the Global South, research has shown that the street acts not only as a 'connector' for people and cars to move but also as a 'container' that holds amidst other everyday practices, the social practices of people (Keswani & Bhagavatula, 2020). In examining the social potential of circulation spaces within neighbourhoods, it is found that even when designed as spaces for efficient movement, such spaces also work for pleasurable walking and informal social activity, especially when they have overlapping usages (Aelbrecht, 2019). However, when there is random movement of people and goods in public spaces, their encounters can be unpredictable (Franck & Stevens, 2006). Aelbrecht (2016) suggests that spaces of circulation may be termed as 'fourth places' when these spaces are characterised by 'in-betweenness' and 'publicness'. While there has been substantial work to understand public spaces and social interactions, the movement of people and goods in relation to time and its influence on the social life of streets has not been sufficiently studied. This chapter addresses this gap by using the lens of Cresswell's (2010) work on mobility, Lefebvre's (2004) ideas of rhythmanalysis and Goffman's (2008) thinking on behaviour in public places to interpret the relation between space, time and movement of people and goods in an Indian marketplace.

The everyday practice of chai delivery becomes the starting point for understanding how marketplace environments work as social spaces by researching how the act of delivering tea to customers sometimes displaces the flow of people or alters their patterns of movement. Chai is seen as a fundamental currency of male friendship, figuring in social rituals enacted at thousands of small shops and stalls in Indian cities (Lutgendorf, 2012). The chapter discusses the aspect of rhythm concerning social interaction, where 'rhythm' is defined as a robust, repeated occurrence of movement patterns. Human geographer Cresswell (2010) suggests that mobility can be thought of as a complex relationship between movement, representation and practice. The chapter draws upon the rhythmanalysis work of the philosopher and sociologist Henri Lefebvre (2004) to understand the relation between social interaction, movement and rhythm (see also Chapter 4). Lefebvre asks us to observe a 'crowd' at peak times and listen to its murmur suggesting that we might find currents and order that emerge from rhythms of accidental and determined encounters in the apparent disorder. One may also be able to analyse this in terms of the work of the social psychologist Ervin Goffman (2008), who defines 'gathering' as that set of two or more individuals who are at the moment in one another's immediate presence.

Research setting: Bhadra Plaza in Ahmedabad

The field investigations were carried out in Bhadra Plaza within the Manek Chowk locality in the old city of Ahmedabad, India (Figure 3.1). The city of

Figure 3.1 A typical chai stall in Manek Chowk, Ahmedabad.

Source: author.

Ahmedabad is divided physically into distinct eastern and western parts due to the Sabarmati river that flows through it. On the east bank of the river is the old city, which includes the central area of Bhadra. The Bhadra Plaza is a public square in front of the Bhadra Fort, a citadel built in 1411. The plaza gets its name from the Bhadra kali temple located within the fort precinct (Sharma et al., 2019). The plaza has always been a commercial hub and marketplace with both formal and informal vending activities. In 2013, a Bhadra Plaza redevelopment project was initiated to enhance the precinct as a public space drawing upon its historical value. However, after the renovation of the plaza had been completed in 2014, the street vendors reappropriated the space. Today, it continues to function as a vibrant marketplace with over 1,200 street vendors occupying it (Colin, 2018).

The area of Manek Chowk within which the Bhadra Plaza is situated has always been a natural market where sellers and buyers have traditionally congregated. As Mehta and Gohil (2013) point out, natural markets in India are not established as part of the city planning process. They tend to emerge from spatial practices that result from socio-economic conditions and are usually characterised by a complex web of interactions. Neethi et al. (2019) suggest that Manek Chowk is not just a commercial space. Instead, they look at it as an entity with physical elements, activities and relationships that contribute meaningfully to placemaking processes. It is within such a marketplace environment that this study on rhythmic encounters is situated.

Spatial ethnography as a methodology

The chapter draws upon the output from a winter school on 'The Everyday City' conducted at CEPT University, Ahmedabad in 2015. The workshop had 15 participants who were architecture students from Aarhus School of Architecture (Denmark), University of Melbourne (Australia) and CEPT University (India). Each group project was done by three students—one each from Denmark, Australia and India. The study discussed in this chapter was a group project on ten chai vendors situated within the Bhadra Plaza. Spatial ethnography was used as the methodology that includes both participant observation and detailed spatial analysis (Gidwani & Chari, 2005; Kawano et al., 2016). Scholars have used this method in different contexts; for example, it has been used in a mapping of street vendors contesting sidewalk space in Ho Chi Minh City (Kim, 2015). Duneier and Carter (1999) apply it in their investigation of street life and political economy in New York City, while Keswani (2019) uses it to understand the antecedents of informality in a marketplace in Bangalore.

For this study, open-ended interviews were conducted with the chai vendors, the delivery agents—referred to in this study as runners—and a few regular customers over a period of two weeks. Interactions with the vendors and runners happened more than once, with the initial conversations being geared towards building familiarity with them. A total of 30 interviews were completed including ten chai vendors and 20 runners across the different tea stalls. The chai vendors and runners were continuously engaged with the making and delivering of the tea and hence, could not be interrupted for in-depth interview sessions at specific times. Instead, the questions were asked at different times as and when an opportunity opened up in the customers' absence. The movement pattern of the chai vendors was understood through participant observation. Sometimes, visits were made to the plaza at 6 a.m. to mark the runners' routines from the start of the day. The paths of the delivery agents of the chai vendors were physically traced by following them. The initial rapport building had helped in making it easier to accompany or follow them. The movement patterns observed were recorded through sketches, which were later translated into representational maps. These maps were further analysed and interpreted along with the observations of the social interactions between chai vendors, runners and customers.

Patterns of movement

In the production of mobility, it has been suggested that physical movement can take place in three ways—people move, things move and ideas move (Cresswell, 2010). In this study, we looked at how the runners move, how the chai moves and how the emptied cups move back to their point of origin. The runners walked from the stalls to the different customers—the formal and informal vendors within the marketplace. The analysis dwells mainly on the movement of the runners, their delivery points and pause points for social interaction. It was found that the static business (i.e. traffic at the push-cart locations) peaked between 10 a.m. and 11.30 a.m., while the dynamic business peaked between 8–9 a.m. and 3–4 p.m. when the runners delivered chai to different parts of the plaza. Each chai vendor had six to eight runners who carried the tea either in an aluminium kettle with six to eight empty glasses or in a flask with 20 smaller cups/glasses.

The tea was dropped off at the regular customers who had placed orders for a chai for the same time every day. Soon after, the runner returned on the same path, collecting the empty glasses on his way back. On the next round, the kettle was filled once more with hot chai, and the same set of actions was repeated.

The study found that the runners did not seem to be a part of any 'gathering' that they served at their delivery points, as they moved from one shop to the next. Goffman (2008) has described the spatial environment within which an entering person becomes a member of an existing gathering as a 'situation'. Here, the runners were found to belong momentarily to such a 'situation' with their level of engrossment being minimal. They did have pause points for social interaction when they met other runners along the way. Once the chai was delivered, it was found that shop owners came together briefly to engage in business talk or engaged in informal conversations with their customers over a cup of chai.

The zones of delivery were about the same in size for most runners and took about 10–15 minutes to traverse. In each run, about 20 cups of tea were sold. In one hour, each runner made four rounds, but from 11 a.m. to 11:30 a.m., it was often six runs in half an hour. In some of the smaller streets, the runner moved in a zigzag manner as he handed over tea glasses to every shop. Porcelain cups were used for the regular customers, which were left behind and collected when the runner returned along the same route after 15–20 minutes. Disposable paper or plastic cups were used when the runner did not want to go back the same route. The runners who had larger territories to cover often stopped by at a restaurant or at a community tap to wash their porcelain cups and continue with their deliveries. One may infer that in the daily life of tea vendors, there may be two kinds of movement: first, when he is stationary and his goods (tea) move, and second, when he moves his business location. For instance, one of the chai vendors was involved in the tea business from 6 a.m. to 12 midnight, after which he moved from the plaza to a neighbouring street where he sold omelettes until 3 a.m.

Rhythmic encounters in the marketplace

There seems to be a broad range of social interactions in the marketplace - a continuum of acts of different persons mutually present to each other (Goffman, 2005). There are encounters between people at different times and at different places that repeat themselves cyclically or linearly and are therefore referred to as rhythmic encounters. The movement patterns of one chai runner are one layer of many pattern layers that exist in the Bhadra Plaza at any given time. The rhythmic encounter layers exist alongside the movement layers, and these layers have a relationship with each other. If one can connect the rhythm, the time when the pattern repeats itself to the movement and where the encounters take place, one has made the connection between movement, time and space. One may then be able to suggest how these spaces might be designed to enhance the encounter or to ensure that it does not hinder other people and objects in the marketplace.

A social gathering can often be a shifting entity (Goffman, 2005). The space and gatherings around the chai stalls were found to change over time, from being empty to partly occupied to completely full. How men and women sat or stood around the stall sometimes differed. It is not enough to create spaces that allow people to come and go; it is just as important to create conditions that enable them to move and linger in these spaces (Gehl, 2011). Some customers came for a quick chai at the Bhadra Plaza and moved away, while others stayed longer. There were chai vendor-customer interactions, customer-customer interactions, customer-runner interactions and runner-runner interactions. In one instance, there was a cigarette shop next to the chai stall, so the smoking and tea drinking became parallel activities. While many customers were shop owners or shoppers, a few were customers who regularly came to a nearby temple and thereafter stopped by for a cup of chai.

In this study, chai becomes the instrument enabling either longer or shorter interactions depending upon the situation and the time of the day. There is a continual interlacing of the daily ritual of conducting business with social interaction, what Mehta (2013) refers to as the 'communal street'. He suggests that such a street is typically characterised by a high level of trust and social capital that the encounters facilitate. In another instance, it is suggested that an emergence of 'situational social capital' can occur where the trust and norms between street vendors only exist during the situation, such as the peanut fair in Bangalore, an annual fair which is both a marketplace and a cultural festival (Keswani & Bhagavatula, 2015). Here, the space itself is a temporal construct, as the peanut fair occupies an arterial road of the city for only two days in a year.

One of the distinctive features of face-to-face interaction has been said to be the richness of information flow and the facilitation of feedback (Goffman, 2008). The interactions emerging from social gatherings around the chai stall are also found to shape themselves according to the presence or absence of shade. The sun's movement through the day affects how and where shadow is generated, and how people sit or stand as they drink their tea. The everyday rhythm of the sun either overlaps or not with the business rhythms (i.e. the business hours of the formal shops) and, therefore, the chai demand cycles. The conversations tend to be longer and more intense in the evenings when the workday is almost coming to an end, while they are shorter and less intense in the mornings.

Based on Goffman's work, Porta and Renne (2005) propose street indicators that can help planners to improve the vitality of a street. They suggest that developing indicators can allow us to quantify the formal components of the design of public spaces. One of these is 'social width', defined as the breadth of the street that affects human interaction across a traffic area, where the restriction on interaction due to traffic lanes is measured. It may be possible to develop an indicator that measures the distance of a chai stall from zones of high footfall, which might impact the customer base and the intensity of rhythmic encounters. In one instance, a chai stall at Bhadra Plaza was located next to a popular fish market that was open for a few hours every morning. This market generated chai customers who were either fish vendors or their customers. Later in the afternoon, this fish market space changed into a shoe market, resulting in a different set of customers. Here, the social width could be measured as the distance between the chai stall and the fish market or the shoe market. Similarly, one might derive an indicator from the location of a chai stall and a temporal clustering of flower vendors outside a temple. These indicators could then be used in the redevelopment of marketplaces in other parts of the city by proposing locations of vendor clusters, both for improving their efficiency and their social environment.

Conclusion

This chapter has attempted to interweave the different (im)mobilities within an Indian marketplace scenario. The case of chai vendors employing runners to deliver tea illustrates the relation between social interaction, time and movement. It illustrates how the movement of one (runner, kettle of chai) results in the immobility of another (people pausing to drink their tea, shop owners not required to close their store to fetch tea). Human mobility has the power of transformation within it, as it can generate different experiences (Cresswell, 2010). At the Bhadra Plaza, both economic and social interactions happen along the paths of movement. The interacting individuals could be members of the same actor group—chai vendors, regular customers or familiar strangers. From his anthropological study in a north-Indian city, Berreman (1972) has suggested that social interaction and structure may comprise complex behavioural choices based on implicit rules. These rules take into account various situations, matters of personal and temporal circumstance, an immediate or ultimate audience for one's behaviour and the actors' own definitions of the social situation and their own role in it.

At the Bhadra Plaza, a chai vendor controls the number of runners, the distance they can cover, but not the time taken to travel this distance. The runners' movements are interspersed with time lapses either at the *delivery points*, which is a variable dependent on the chai demand, or at the *pause points*, which may be random, informal conversations between fellow runners or between runners and other street vendors (Figure 3.2). At the delivery point, the runner acts as a

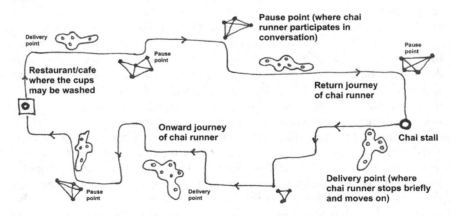

Figure 3.2 The relationship between space, time and movement.

Source: author.

catalyst because he brings the chai that becomes a reason for others to relax and make conversations among themselves. He may not participate in these conversations if he finds himself outside the social structure. Through this analysis, one can establish a relationship between places of social interaction, time and movement.

It is possible that if the chai vendor did not employ runners, the pattern of movement would differ, that a greater number of shopkeepers would walk to the chai vendor and his vending space. This extra footfall might result in a higher density of human mobility at certain times of the day when customers are moving through the streets to fetch their chai. Alternatively, it is also possible that there would be a higher number of chai vendors located at closer intervals to reduce the walking distance for the shop owner—who is a customer—so that he does not have to close his shop while he fetches his tea. The nodes of interaction or spaces of conversation would thus be located differently within the marketplace, and the connectors between them would be different. One might infer from this analysis that the movement of the chai runners helps understand biological, psychological and social rhythms. The chai is a daily habit associated with the biological rhythm of thirst, which relates to at least two specific times of the day 10.30–11.00 a.m. and 3.30–4.00 p.m. It is also a social gesture as people pause work and make conversations over the cup of chai. Hence, it is a rhythm of the body as well as of society.

Lefebvre (2004) has analysed the difference between linear and cyclical time, and the contrast between clock time and lived time. For him, both time and space are lived experiences. He suggests that just as cartesian geometry is a reductive way of understanding space, so too is the measure of time, 'the clock', a reductive comprehension. Here, in the marketplace, one observes a cyclical repetition or rhythm that originates in nature: days, nights and seasons and a linear repetition that comes from social practice and human activity: actions and movements. One finds that there are cyclical rhythms that last for a limited period and then restart. When dawn arrives, the marketplace resets itself, inaugurating the return of the *everyday* yet again. As the spatial patterns change due to a shifting of how people and goods are positioned and due to a change in movement patterns of people and goods that are dynamic, the relation between time and space is also found to change.

The findings in this study also allow us to bring in the idea of time measure. On the one hand, there is a specific time for a chai; and on the other hand, there is the time gap between the morning chai and the afternoon chai. One can generate a measure for time and also a measure for space which may be the distance between the chai vendor's location and the shop's location where the customer is situated. If there is a push-pull exchange between the multiple rhythms, it either accelerates or hinders the movement of people and goods. For instance, when it rains and the weather gets chilly, the demand for hot cups of chai goes up, and the frequency of runner trips increases. Sometimes, the pace or speed of movement is altered depending upon the urgency of both the regular and the infrequent customers. When there is an interaction between a place, a time and an expenditure of energy or movement, there is rhythm.

There are different urban design principles that emerge from this study. First, there are pause points being generated along the everyday routes of the chai runners that could be designed as resting places or those small, public spaces for gatherings to situate themselves. The second principle that emerges is from knowing 'rhythms' or knowing how often an activity repeats itself in a day, when exactly it occurs and the duration for which it lasts. The temporal dimension and its understanding make it possible to use the same street space or same open marketplace for different activities at different times of the day. Third, the design intent may be to make changes in the space in order to alter the rhythmic encounter itself. If one were to introduce either the 'shade of a tree' as an urban design element or a seating arrangement at select points, one may be able to enhance or endorse the rhythmic encounters between people. Fourth, one may borrow from Lynch's (1972) idea of 'change intelligence' to create a public, temporal model of the marketplace that could expose citizens to the diversity of people, activities and forms in public spaces. Further, if rhythms are understood across marketplaces in the city, they can be incorporated into future street design guidelines and urban street vendor policies.

Acknowledgements

I would like to thank all the participants of the CEPT winter school on *The Everyday City* held in 2015 for the sharing of ideas and insights during the workshop. In particular, I would like to thank Parisa Abuhamzeh, Srinavya Annem and Will Rathgeber for their fieldwork and insights on the chai vendors. I am grateful to Prof. A. Srivathsan, Prof. Yatin Pandya and Prof. Thomas Hilberth for their feedback during the workshop review sessions.

References

Aelbrecht, P. (2016), 'Fourth places': The contemporary public settings for informal social interaction among strangers. *Journal of Urban Design*, 21(1), 124–152.

Aelbrecht, P. (2019), New public spaces of circulation, consumption and recreation and their scope for informal social interaction and cohesion. In: Aelbrecht, P. & Stevens, Q. (Eds.), *Public Space Design and Social Cohesion: An International Comparison* (pp. 199–219). New York: Routledge.

Berreman, G.D. (1972), Social categories and social interaction in urban India. *American Anthropologist*, 74(3), 567–586.

Colin, L.O. (2018), Street vending from the right to the city approach. The appropriation of Bhadra Plaza. In: Cabannes, Y., Douglass, M. & Padawangi, R. (Eds.), *Cities in Asia by and for the People* (pp. 259–282). Amsterdam: Amsterdam University Press.

Cresswell, T. (2010), Towards a politics of mobility. *Environment and planning D: Society and Space*, 28(1), 17–31.

Duneier, M. & Carter, O. (1999), *Sidewalk*. New York: Farrar, Straus & Giroux.

Franck, K. & Stevens, Q. (Eds.) (2006), *Loose Space: Possibility and Diversity in Urban Life*. New York: Routledge.

Gehl, J. (2011), *Life Between Buildings: Using Public Space*. London: Island Press.

Gidwani, V.K. & Chari, S. (2005), Grounds for a spatial ethnography of labour. *Ethnography*, 6, 267–281.

Goffman, E. (2005), *Interaction Ritual: Essays in Face-to-Face Behavior*. New Brunswick: Transaction Publishers.

Goffman, E. (2008), *Behavior in Public Spaces*. New York: The Free Press.

Janssens, F. & Sezer, C. (2013), Marketplaces as an urban development strategy. *Built Environment*, 39(2), 169–171.

Kawano, Y., Munaim, A., Goto, J., Shobugawa, Y. & Naito, M. (2016), Sensing space: Augmenting scientific data with spatial ethnography. *Geo Humanities*, 2(2), 485–508.

Keswani, K. (2019), The logic of design: Its role in understanding the antecedents of urban informality. *Journal of Urban Design*, 24(4), 656–675.

Keswani, K. & Bhagavatula, S. (2015), Territoriality in urban space: The case of a periodic marketplace in Bangalore. In Evers, C. & Seale, K. (Eds.), *Informal Urban Street Markets* (pp. 152–162). New York: Routledge.

Keswani, K. & Bhagavatula, S. (2020), The ordinary city and the extraordinary city: The challenges of planning for the everyday. *APU Working Paper Series*. Bengaluru: Azim Premji University.

Kim, A.M. (2015), *Sidewalk City: Remapping Public Space in Ho Chi Minh City*. Chicago: University of Chicago Press.

Lefebvre, H. (2004), *Rhythmanalysis: Space, Time and Everyday Life*. London: Continuum.

Lutgendorf, P. (2012), Making tea in India: Chai, capitalism, culture. *Thesis Eleven*, 113(1), 11–31.

Lynch, K. (1972), *What Time Is This Place?*. Cambridge, MA: MIT Press.

Mehta, R. & Gohil, C. (2013), Design for natural markets: Accommodating the informal. *Built Environment*, 39(2), 277–296.

Mehta, V. (2013), *The Street: A Quintessential Social Public Space*. London: Routledge.

Narayan, R.K. (1989), *A Story-Teller's World*. New Delhi: Penguin Books India.

Neethi, P., Kamath, A. & Paul, A.M. (2019), Everyday place making through social capital among street vendors at Manek Chowk, Gujarat, India. *Space and Culture*, doi.10.1177/1206331219830079.

Porta, S. & Renne, J.L. (2005), Linking urban design to sustainability: Formal indicators of social urban sustainability field research in Perth, Western Australia. *Urban Design International*, 10(1), 51–64.

Sharma, U., Mistry, P. & Prajapati, R. (2019), Revitalization strategy for historic core of Ahmedabad. *International Journal of Environmental Science & Sustainable Development*, 4(2), 45–60.

4 Spectral analysis of rhythms in urban marketplaces

A day in Esat Marketplace of Ankara (Turkey)

Nihan Oya Memlük-Çobanoğlu and Bilge Beril Kapusuz-Balcı

Introduction

The study of urban public spaces through unveiling the public life that takes place in them has been an ongoing quest after the pioneers Jacobs (1961), Appleyard (1980), Whyte (1980) and Gehl (1987). The investigations on public life in public space, in other words how life unfolds in city space (Gehl & Svarre, 2013), altered our understanding of production of space more as a collaborative act and resulted in a growing number of studies centred around observing, listening and asking questions to the users of space rather than the formal approaches of the 20th century (Project for Public Spaces, 2007). In return, these studies helped to prevent the failures and mismatches between the needs and aspirations of the end-users with the affordances of the urban built environment (Thomas, 2016).

Rhythmanalysis, coined by Lefebvre (2004), provides a substantial perspective on depicting the intrinsic, lived-in nature (Lefebvre, 1974) of public life in public space. It allows for an elaborated comprehension of everyday life (Crang, 2001) by revealing rhythmic differences in places (Koch & Sand, 2010). It also puts forth the relational and inter-scalar aspects of social space and time (Reid-Musson, 2018), tracks and demonstrates synchronic practices and temporal patterns (Edensor, 2010a) and depicts the timespace routines in public space, which allows for a better narration of public life (Mulíček et al., 2015). Rhythmanalysis also enables an understanding of the production of space as a collaborative process, a 'becoming' with emergent possibilities (Edensor, 2010b). Providing an essential perspective for diverse fields such as architecture, urban planning and social sciences (Koch & Sand, 2010), there has been a widespread focus on rhythmanalysis studies, especially after the translated publication of rhythmanalysis (Lefebvre, 2004) in English. These studies include empirical research, quests for methodologies and discussions on the sufficiency of the theory.

Marketplaces are unique public spaces where diverse rhythms overlap, producing a rich and vibrant place-temporality and public life (see also Chapter 3). This chapter on marketplaces aims to understand and unveil the idiosyncratic rhythmic complexity of the marketplace. It proposes 'spectral analysis' as a coherent method to represent and communicate the rhythms of public spaces through its tools for critical observation based on visual documentation and interpretation

DOI: 10.4324/9781003197058-4

of a spectrum of rhythms. Spectral analysis is conducted in Esat Marketplace with a rhythmanalytical perspective for the critical assessment of its recent spatial transformation in 2018. As such, this chapter aims to provide insights to inform design practices for marketplaces.

Rhythmanalysis in marketplaces

In his seminal book, Lefebvre (2004) defines rhythm as the overlapping of "*a place, a time and an expenditure of energy*" (Lefebvre, 2004: 15) and "*a mode of analysis rather than just an object of it*" (Elden, 2004: xii). The main attributes intrinsic to rhythm are defined as repetition (either cyclical as in nature or linear produced by social practices), measure and movement (Lefebvre & Régulier, 2004). In this regard, Lefevbre's rhythmanalysis provides a substantial perspective on linking space and time through "*a temporal understanding of place*" (Edensor, 2010b: 1–2) which is referred to as "*a localised time, or ... a temporalised space*" (Lefebvre & Régulier, 2004: 89).

According to Lefebvre (2004), there are three types of rhythms: natural, corporeal and mechanistic. The natural rhythms of cyclic repetition, the physiological cycles of the body and the mechanistic repetition of spaces are superimposed in space, forming a complex 'polyrhythmic ensemble' (Lefebvre, 2004). Accordingly, the frequency, intensity and duration of the rhythmic clustering in a timespace resulting in a typical place flow is termed 'place-temporality' (Wunderlich, 2013). These various rhythms overlapping, interfering and interacting with one another continuously and iteratively are also open for contingencies, asynchronous moments for creating difference and variety at each instance. Lefebvre (2004) also mentions different types of rhythmic states: polyrhythmia is an ensemble of multiple rhythms; eurhythmia refers to the normal state of everydayness; arrhythmia is the times of disturbances to eurhythmia.

In terms of rhythmic agents, not only human but also the non-human agents—the material constituents of the surrounding—and their rhythms are set as the generative forces of place-temporality. Chen (2013) defines the role of the non-human agents as timing and spacing practices in everyday settings and the relation between bodily rhythms and the material surrounding as an ongoing negotiation. Wunderlich (2013: 383) emphasises this negotiation as follows"

> In an urban place, the patterns of people's movements, encounters, and rest, recurrently negotiating with natural cycles and architectural patterns, merge into expressive bundles of rhythms which give a place its temporal distinctiveness.

None of the rhythmic agents can be defined as the sole determiner of the place-rhythm. The variety of and relations between rhythms produce the complexity of everyday settings, the polyrhythmia. Chen (2013: 535, 546) refers to this polyrhythmic state as a "*dynamic meshwork*" in which there is an ongoing "*transmission, interaction and exchange*" between rhythms. Consequently, this polyrhythmic state, the harmony and contradiction between rhythms, animates public space (Lefebvre, 2004; Mulíček et al., 2015) and produces the unique place-temporality. In return, the unique place-temporalities contribute to the sense of place (Crang, 2001), the characteristics (Wunderlich, 2008; Edensor, 2010a) and the identity (Mulíček et al., 2015) of localities.

Marketplaces with their flexible spatial and temporal organisations (Schappo & Van Melik, 2017) and vibrant and inclusive characters (Janssens & Sezer, 2013) have unique place-temporalities produced by various rhythmic agents. Seamon and Nordin (1980) compare the unique place-temporality of marketplaces to a 'space ballet' based on the regular timespace routines performed by habitual bodily rhythms and unexpected contingent occurrences between diverse bodily rhythms. These intrinsic qualities make marketplaces an ideal subject for rhythmanalysis.

Narrating the marketplace: Spectral analysis of rhythms

Spectrum originally comes from its early 17th-century use of spectre referring to the French *spectre* 'an image, figure, ghost' and the Latin word *spectrum* as 'an appearance, image, apparition, spectre', having its roots in the *specere* 'to look at, view'. In the 1670s, spectrum was used to describe the "*visible band showing the successive colours, formed from a beam of light passed through a prism*" when the image of the band was first recorded visually (Online Etymology Dictionary, 2001).

Spectrum is a metaphoric phenomenon for this inquiry in defining both the field and the research methodology concerning the word's etymological origins. Accordingly, the marketplace, as a field of rhythms, presents a spectrum of flows, movements and relations of coexisting and encountering bodies, objects and spaces. Spectral analysis methodologically refers to an act of deeper 'looking at' a place through a broader observation of its diverse rhythms by the act of on-site observations and visual recording and narration of place-temporality. This study argues that rhythmanalysis lacks an analytical and representational inquiry method into the 'spectrum of rhythms' at the polyrhythmic state of public space. It offers 'spectral analysis' as a methodological contribution.

Koch and Sand (2010) claim that the methods and the means of documentation (photos, video and performances) in rhythmanalysis studies need advancements. Simpson (2012) proposes time-lapse photography as a complementary tool capable of capturing the complex durational unfolding of events forming the polyrhythmic ensemble and the emergent occurrences. Time-lapse photography enables both recording and communicating a multitude of durations in a series of consequential images of the place. It represents the ensemble created by human and non-human bodies that coexist and correlate in a rhythmic condition. Yet, time-lapse photography falls short in terms of depicting the tactile experience of space. Thus, additional qualitative data collection methods are employed during the research, including on-site observations, researchers notes and photographic survey (instant photographs).

In this context, a complete cycle of Esat Marketplace was observed, experienced and recorded from Saturday night when the marketplace is set up till the hours after its closing on Sunday. The time unit referred to as 'a full cycle' for observing the marketplace, manifests a transitive interval that interlocks different spatio-temporalities of this locale. The focus on the complete cycle is related with the dynamic rhythmicity of this period with the use of space as a marketplace. This space is otherwise used almost in an obsolescent state as a parking area by the residents. Although it is occasionally used as a playscape by the younger population and for certain events (concerts, charity bazaar, etc.), it is on an irregular and limited basis. It is important to note that more longitudinal observations could provide a broader perspective, such as seasonal differences. However, the focus is on the internal rhythms of the marketplace within the scope of this study. The on-site observations were conducted through multiple visits during July 2020 (July is selected to omit constraints related to bad weather conditions both for the users and documentation). The time-lapse photographs are taken during a Sunday (July 2020) at eight-time intervals (21:00 (Saturday), 06:30, 09:30, 12:30, 15:30, 18:30, 21:30 and 23:00) during 10-minute sessions, which refer to the specific time intervals with characteristic place-temporalities. Observations were conducted both from 'above site' and 'on-site', echoing the characteristics of the 'rhythmanalyst' defined by Lefebvre (2004).

A day in Esat Marketplace

Esat Marketplace is located in the Küçükesat district of Ankara, still preserving its neighbourhood culture, unlike other neighbourhoods that have faced disruptive urban transformation. The marketplace is located at the core of the district, with its main entrance located at the intersection of two main arterial roads, while there are two other entrances on the adjacent streets. As a long-standing and vibrant constituent of the district, Esat Marketplace is selected to observe public life for its unique spatio-temporality, significantly changed due to a transformation project by the local municipality in 2018.

Still being a vibrant public space, the spatial transformation led by the local decision-makers brought the marketplace up for a critical discussion. The former configuration of the marketplace connected two street levels with a gradual spatial organisation, whereas the recently constructed concrete slab, which accommodates a park at the higher street level and becomes the roof of the marketplace on the lower level, separates the two levels both physically and functionally. Although every Sunday, when the market is set, the two layers of public space with two different uses—commercial and recreational—overlap, they largely operate independently.

As Edensor (2010b: 1) denotes, time should be "*conceived as dynamic, multiple and heterogeneous*" rather than "*a singular or uniform social time stretching across a uniform social space*". Relatedly, time is approached as 'social time' produced through polyrhythms of the marketplace in this analysis rather than the linear conception of time as in clock time. Thus, cyclic temporal intervals, each having its characteristic place-temporality, are referred to during the analysis. Polyrhythmic order (Lefebvre, 2004) requires conceptualisation, observation and documentation of both the 'regular' and the 'unexpected' patterns (Seamon & Nordin, 1980). This study examines the rhythms in Esat Marketplace in terms of both regular and emergent patterns.

Regular rhythms: Eurythmic state of the marketplace

The regularly patterned rhythms overlap in the marketplace and constitute the eurythmic state (Lefebvre, 2004), the routine cycle of Esat Marketplace. This cycle starts on Saturday night when the sellers bring their stalls and durable products. In this interval, the most visible rhythmic agent is the stallholders placing their equipment, contributing to the 'production of space' (Lefebvre, 1974) through their corporeal rhythms. As for non-human agents, the food trucks, on a more temporal basis as they enter and leave the site, are also at the forefront by "*creating mobile flows of varying in tempo, pace and regularity*" which in return "*contribute to the spatio-temporal character of place*" (Edensor, 2010b: 5). Insomuch as, the structural order of the columns operates in the organisational placement of the stalls, which illustrates an example of the non-human agents in timing and spacing spatial practices. At this interval, rather than a dominating relation, rhythms coexist within a harmony.

The early morning is the interval when the sellers complete the setup of the marketplace. Sellers are again the dominant rhythmic agents with their corporeal rhythms, while the semi-fixed materials such as umbrellas and food boxes are in a constant flux creating a flow. For the rhythmanalysts 'on-site', it is possible to encounter the social interactions between strolling sellers as they greet each other and chat. In this way, the sellers engage in more temporal types of interaction just before they partake in the mechanistic rhythm of the marketplace. Similar interactions occur during the day between the sellers and the regular customers. These social interactions repeating in a cyclic pattern allow for social ties that enhance a sense of belonging—place attachment—towards the marketplace.

During the morning, there is a tranquil atmosphere dominating the marketplace. The flow of buyers, pedestrian movement in and out of the marketplace and stalling cars along the adjacent roads are at a low frequency and tempo than the other day sections. The naturalistic rhythms, the sun and shade, are essential rhythmic agents in this interval. The rhythms of the semi-fixed materials, umbrellas (frequently adjusted to the angle of the sunlight creating their own rhythm), are synchronised with these natural rhythms. Although this rhythm of umbrellas creates a visual dynamic for the above-site rhythmanalyst, it reveals the inadequacy of the affordances of the built environment (Gibson, 1977) in terms of harmonising with the natural rhythms.

In the mid-day, the frequency and movement of buyers and the flow of vehicles often stalling at the adjacent roads increases to a medium level. At this interval, there is a time section when food is served to the sellers from a single position called the 'restaurant'. As such, the sellers can deviate from the dominant hectic rhythm of the marketplace for a small time period. The tea seller has a similar function and contributes to this polyrhythmic ensemble with its temporary and immobile presence (see Chapter 10). This need for both sellers' and buyers' food and drinks produces contingent rhythms through spatio-temporal overflows. Such rhythms include stalls, cabinets of tea sellers, the 'restaurant' and street vendors overflowing into the sidewalk of the adjacent streets and the impermeable walls used for leaning umbrellas, eating on and piling up empty food boxes. Besides, these spatial overflows indicating the inadequacy of the affordances of the built environment prohibit the use of the already narrow sidewalk, resulting in an interfering rhythm.

The secondary entrance on the adjacent street is preferred more commonly at this interval since it is on the same level as the sidewalk. In contrast, the staired main entrance is often preferred by the more temporary buyers with their stalling cars on the roadside. This preference implies a discordance with the intended hierarchy of entrances by design. Similar discordances emerge when undesignated areas are used for corporeal rhythms in various moments. The parapet wall of the stairs becomes a temporary resting spot for the buyers, the shadow of the cabinets and the parapet wall provides a resting space for the street vendors or a porter (*hamal*) sits for a cigarette break on his trolley. These discordances refer to a contradiction between user needs and design that necessitates spatial and temporal 'negotiations' through alternative uses of space.

The afternoon is the high time of the marketplace when all the rhythmic agents, human and non-human, are present on the scene producing a diversity of rhythms with high frequencies and tempos. There is a busy atmosphere dominating the marketplace in which all rhythmic agents in a variety of types of presence synchronise and/or conflict with each other. The varying rhythmic agents and diverse types of presences intensifying in a delimited locale animate the space and promote vitality, while simultaneously creating a busy ambience and congestion. In addition to the contribution of the after-work hours, this intensification is also related to economic (the prices are lowered almost three times during the afternoon to sell all the left products) and natural (cooler temperature of the afternoon) rhythms. The lower prices in the afternoon are referred to as the 'night bazaar' by the residents.

Finally, the night interval begins with the closing section of the marketplace when the sellers dismantle their stalls. In this interval, the frequency and tempo of the rhythms of both human and non-human agents are lowered with respect to earlier sections and leftover food pickers as new human rhythmic agents enter the scene creating another regular pattern. Hence, it can be claimed that marketplaces have their own economic rhythms that contribute to their place-temporality, as discussed previously.

All in all, iterative harmonic and/or colliding coexistence of rhythms are regularly patterned along the day and cyclically repeated weekly, producing the eurythmic state of the marketplace. The diversity and intensity of agents and their variety of modes of presence enrich both the sense of place and time in Esat Marketplace. Yet, there is no single dominant agent or rhythm, but there is a negotiation between rhythms transforming the place-temporality of the marketplace with respect to each time interval (Figure 4.1).

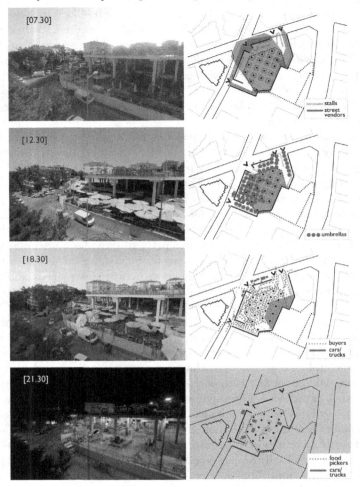

Figure 4.1 Analysis of the rhythms of Esat Marketplace, 2020.
Source: authors.

Emergent rhythms: Moments of arrhythmia

The repetition of rhythms is also open to difference, they can *"interrupt, change, break, or contradict each other or the complex polyrhythms"* (Koch & Sand, 2010: 61–62). Place can only be *"depicted, performed and sensed through its ensemble of normative and counter rhythms"* (Edensor, 2010b: 4). The marketplace, having its regularly patterned rhythms, is also open to differing and spontaneous rhythms. These emergent rhythms unfold through the complex relations between multiple agents, creating the arrhythmic moments when the routine is interrupted. Referring to Edensor (2010b), unravelling the ensemble of normative rhythms and their counterparts requires the participation of the rhythmanalysts in the spatial experience. Therefore, the rhythmanalysts' repeated on-site presence in the marketplace was of great importance for detecting the moments of arrhythmia during the analysis.

In Esat Marketplace, enactment of arrhythmia is related with the intensification and diversification of both the rhythmic agents—their types of presence—and the affordances of the environment. For example, when the staircase is used as a playscape by children, the spatial agency of the stairs creates a different rhythm through the act of playing as an alternative type of presence. In other words, material agents set the affordances for corporeal rhythms while they can be interrupted at these moments of arrhythmia through emergent corporeal rhythms. This arrhythmic moment is related to the recent design of the marketplace since the playground on the upper level is not accessible for the users of the marketplace, as the two levels are not connected.

Another moment of arrhythmia emerges when the sellers create moments of attraction for the buyers through animative performances, such as announcing their goods and prices by singing and shouting, displacing goods by passing from hand to hand in a theatrical way or offering buyers samples for tasting during the day. Thus, their corporeal rhythms disrupt the mechanistic rhythm of the marketplace as well as animate the space. Consequently, all these emergent place-rhythms are important instants that enrich the spatio-temporal character of the place. At the same time, they are deeply interrelated with the affordances of the marketplace, which is subject to critical re-evaluation within this research.

Conclusion

The main conclusion of this research is that the complex polyrhythmic ensemble of regular and emergent rhythms should be investigated in terms of synchronisations and contradictions to understand the true nature of public life in public space. For instance, the corporeal rhythms of the non-human agents, such as the umbrellas, cannot be explored separately from the naturalistic rhythms of the sunlight. The emergence of arrhythmic moments in the afternoon cannot be dissociated from the intensification of human agents and their rhythms increasing in terms of frequency and tempo in the case of Esat Marketplace.

Rhythmanalysis provided the basis for the understanding of the spatio-temporal characteristics and the public life in Esat Marketplace, while 'spectral analysis' enabled the observation and communication of the rhythms created by diverse rhythmic agents. By means of the on-site observations and the time-lapse video created from the above site (see QR code), this method of analysis is used for the narration and spectral representation of the temporal presence of rhythmic agents in a constant flux and in relation to each other. Vergunst (2010) claims that the modes of presence and the internal rhythms of the rhythmic agents also contribute to place-temporality. In a similar vein, all the human and non-human agents and their presence in terms of the regular and the emergent patterns are observed through spectral analysis in Esat Marketplace. These patterns are conceptually described based on their spatial (from fixed to mobile) and temporal (from durable to temporary) characteristics in Figure 4.2. This conceptual figure represents a full spectrum of rhythmic agents and how diversely they relate and contribute to the production of space in Esat Marketplace. The figure demonstrates that the fixed and durable rhythmic agents set the stage for regular patterns, while more mobile and temporary agents create a space for both 'contingent' and 'emergent' patterns. Although emergent patterns, the moments of arrhythmia, seem to be discordances or disruptions from the eurythmic state, they have a '*resultant revolutionary potential within*' (Simpson, 2008: 823) in terms of transforming the mechanistic rhythms. Moments of arrhythmia lie at the heart of unique place-temporalities of public spaces and offer opportunities for vivid and emergent public life.

The interrelations, especially in terms of conflicts and contingencies, between rhythms of the marketplace revealed significant insights for design. The contingent rhythms include the spatial overflow on the entrances and towards the sidewalks or additional umbrellas used for shadowing. These connote to a discordance between the physical affordances of the newly transformed marketplace in terms of meeting the end user's needs. In other words, if space is not designed for the needs of a variety of agents with diverging types of presence, a conflict between rhythmic agents occurs and contingent rhythms emerge in the form of spatio-temporal negotiations. Sletto and Palmer (2017: 2361) define such relations as "*a creative act of negotiating territory*". Such discordances between corporeal rhythms of human and the non-human agents (material surrounding)

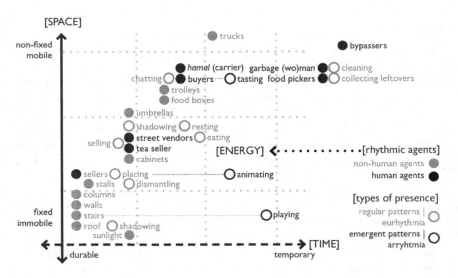

Figure 4.2 Rhythmic agents and their types of presence in Esat Marketplace.
Source: authors.

are referred to by Wunderlich (2013: 407) also as *"opportunities for practical applications of rhythmical analysis to inform design"*. The discordances in Esat Marketplace are related also to the production of space seen solely as a formal, top-down design act amid the reconstruction of the space lacking an 'observational' or 'collaborative' basis. This finding presents important implications for the future design practices of marketplaces.

Approaching the public space on a temporal basis revealed that, in Esat Marketplace, the routine cycle of the space comprises certain intervals with unique spatio-temporal characteristics. These temporal intervals, and the intensity and tempo of diverse rhythms in each interval, necessitate different 'negotiations' between agents. Hence, a temporal understanding of place provides a broader perspective than the conventional approaches privileging solely the spatial dimensions of place, especially for the disciplines related to public space design. It is not sufficient to refer only to a place's unique polyrhythmia or arrhythmic moments but also their values and limitations in terms of creating an 'ensemble' should be analysed to attain more responsive design of the public space.

Visual reference

'A Day in the Esat Marketplace' time-lapse video can be accessed from the link: https://www.youtube.com/watch?v=9udfsJX6DSQ and also from the QR code (Figure 4.3).

Figure 4.3 QR code.

References

Appleyard, D. (1980), Livable streets: Protected neighbourhoods? *The ANNALS of the American Academy of Political and Social Science*, 451(1), 106–117.

Chen, Y. (2013), 'Walking with': A rhythmanalysis of London's East end. *Culture Unbound: Journal of Current Cultural Research*, 5(4), 531–549.

Crang, M. (2001), Rhythms of the city: Temporalised space and motion. In: May, J. & Thrift, N. (Eds.), *Time Space: Geographies of Temporality* (pp. 187–207). London: Routledge.

Edensor, T. (2010a), Walking in rhythms: Place, regulation, style and the flow of experience, *Visual Studies*, 25(1), 69–79.

Edensor, T. (2010b), Introduction: Thinking about rhythm and space. In: Edensor, T. (Ed.), *Geographies of Rhythm: Nature, Place, Mobilities and Bodies* (pp. 1–18). London: Routledge.

Elden, S. (2004), Rhythmanalysis: An introduction. In: Lefebvre, H. (Ed.), *Rhythmanalysis: Space, Time and Everyday Life*. Trans. Stuart Elden, Gerald Moore, (pp. 7–15). London, New York: Continuum.

Gehl, J. (1987), *Life Between Buildings—Using Public Space*. New York: Van Nostrand Reinhold.

Gehl, J. & Svarre, B. (2013), *How to Study Public Life*. Washington, DC: Island Press.

Gibson, J.J. (1977), The theory of affordances. In: Shaw, R. & J. Bransford (Eds.), *Perceiving, Acting and Knowing: Toward an Ecological Psychology* (pp. 67–82). Hillsdale, New Jersey: Erlbaum.

Jacobs, J. (1961), *Death and Life of Great American Cities*. New York: Random House.

Janssens, F. & Sezer, C. (2013). Marketplaces as an urban development strategy. *Built Environment*. 39(2), 168–177.

Koch, D. & Sand, M. (2010), Rhythmanalysis: Rhythm as mode, methods and theory for analysing urban complexity. In: Aboutorabi, M. & Wesener, A. (Eds.), *Urban Design Research: Method and Application* (pp. 61–72). Birmingham: Birmingham City University.

Lefebvre, H. (1974), *Production of Space*. London: Continuum.

Lefebvre, H. (2004), *Rhythmanalysis: Space, Time and Everyday Life*. London: Continuum.

Lefebvre, H. & Régulier, C. (2004), The rhythmanalytical project. In: Lefebvre, H. (Ed.), *Rhythmanalysis: Space, Time and Everyday Life* (pp. 71–83). London: Continuum.

Mulíček, O., Osman, R. & Seidenglanz, D. (2015), Urban rhythms: A chronotopic approach to urban timespace. *Time & Society*, 24(3), 304–325.

Online Etymology Dictionary (2001), *Spectrum*. [Online]. Available at: https://www.etymonline.com/search?q=spectrum [Accessed 1 August 2020].

Project for Public Spaces (2007), *What Is Placemaking?* [Online]. Available at: https://www.pps.org/article/what-is-placemaking [Accessed 20 January 2021].

Reid-Musson, E. (2018), Intersectional rhythmanalysis: Power, rhythm, and everyday life. *Progress in Human Geography*, 42(6), 1–17.

Schappo, P. & Van Melik, R. (2017), Meeting on the marketplace: On the integrative potential of The Hague Market. *Journal of Urbanism*, 10(3), 318–332.

Seamon, D. & Nordin, C. (1980), Marketplace as place ballet. *Landscape*, 24, 35–41.

Simpson, P. (2008), Chronic everyday life: Rhythmanalysing street performance. *Social & Cultural Geography*, 9(7), 807–829.

Simpson, P. (2012), Apprehending everyday rhythms: Rhythmanalysis, time-lapse photography, and the space-times of street performance. *Cultural Geographies*, 19(4), 423–445.

Sletto, B. & Palmer, J. (2017), The liminality of open space and rhythms of the everyday in Jallah Town, Monrovia, Liberia. *Urban Studies*, 54(10), 2360–2375.

Thomas, D. (2016), *Placemaking: An Urban Design Methodology*. New York: Routledge.

Vergunst, J. (2010), Rhythms of walking: History and presence in a city street. *Space and Culture*, 13(4), 376–388.

Whyte, W.H. (1980), *The Social Life of Small Urban Spaces*. New York: Project for Public Spaces.

Wunderlich, F.M. (2008), Symphonies of urban places: Urban rhythms as traces of time in space. A study of 'urban rhythms'. *KOHT ja PAIK/PLACE and LOCATION Studies in Environmental Aesthetics and Semiotics VI* (pp. 91–111), Estonia.

Wunderlich, F.M. (2013), Place-temporality and urban place-rhythms in urban analysis and design: An aesthetic akin to music. *Journal of Urban Design*, 18(3), 383–408.

5 Adaptable market-making in eThekwini

Exploring practices of street trading in a South African urban space

Noxolo Ndaba and Karina Landman

Introduction

Despite the importance of informal trading to offer sustainable livelihood opportunities to millions of people in South Africa, tensions remain between traders and city council officials. These tensions revolve around visions for orderly and clean public spaces versus the need for efficient and accessible places to trade. Sanitised space refers to processes in which urban space is 'cleansed' from all things that make it dirty, undesirable and differentiated (Smith & Walters, 2018). However, drawing from civic humanism, public spaces should also be considered in terms of politics, rights and social justice. Great public spaces and sidewalks in particular need to be tolerant and inclusionary (Blomley, 2010: 11). Warwick Junction in the City of eThekwini (Durban) is one of the biggest transport hubs in the country with popular market spaces featuring large numbers of informal traders. These tensions were particularly evident during the upgrading of the Warwick Junction and resulted in many changes, some planned and others not. This raises some questions regarding the change of public spaces and the impact on the opportunities for informal trading in these spaces.

As part of ongoing work on the transformation of public space in South Africa, we investigated the nature and use of streets in eThekwini. We also explored the experiences and perceptions of the users of these spaces. Although numerous factors were mentioned, we were amazed by the stories and experiences of the traders, some of whom have been there for many years, and by various socio-spatial transformations evident in the city.

In this chapter, we examine the trading activities performed in Victoria Street and how these are embodied and experienced, particularly in light of the need for sustainable livelihood opportunities in public space (Chen, 2020) and the relation of resilience to urban sustainability (Peres et al., 2017). The chapter argues that it is the adaptive capacity of the socio-material relations in Warwick Junction, facilitated by the socio-spatial and institutional resilience of the traders, trading spaces and planners, which allow the users of the public spaces to benefit from these marketplaces in various ways. The analysis is part of an ongoing conversation about the value of repeatedly unsanctioned or undesirable activities such as informal trading or marketplaces to challenge the status quo,

DOI: 10.4324/9781003197058-5

bring new meaning to public space and facilitate economic opportunities (Lyndon & Garcia, 2015; Madanipour, 2017; Landman, 2019; Hou, 2020).

Reconsidering the notion of adaptive capacity within informality

Resilience refers to the ability of a system to absorb disturbance and retain its basic function and structure. Therefore, it relates to concepts of sustainability and the challenge of servicing current service demands without eroding the potential to meet future needs (Walker and Salt, 2006, 1–2). Given this,

> [u]rban resilience refers to the ability of an urban system—and all its constituent socio-ecological and socio-technical networks across temporal and spatial scales—to maintain or rapidly return to desired functions in the face of a disturbance, to adapt to change, and to quickly transform systems that limit current or future adaptive capacity.
>
> (Meerow et al. 2016: 45)

Urban resilience has many dimensions, including social, spatial and institutional dimensions (Masnavi et al., 2019). Social resilience is concerned with the ability of communities to sustain themselves in the aftermath of shocks and stresses. According to Keck and Sakdapolrak (2013: 4), social resilience comprises of coping capacity or the ability of social actors to cope with and overcome all kinds of adversities. This includes both adaptive capacity—the ability to learn from experiences and adjust to future challenges in people's everyday lives—and transformative capacity, which refers to the ability to construct sets of institutions that foster individual welfare and sustainable societal robustness towards future crises. Researchers of complex adaptive systems describe social resilience as the adaptive and learning capacity of individuals, groups and institutions to self-organise in a way that maintains system function in the face of change or response to a disturbance (McLean et al., 2016: 33).

Spatial or physical resilience can be understood as the robustness of urban structures and networks against random failures (Salat & Bourdic, 2012: 65) or changes and is influenced by the urban form and its relation to land use. A few studies have started to investigate spatial resilience at a neighbourhood level (Nel & Landman, 2015; Feliciotti et al., 2017) or related to public space (Landman & Nel, 2021). Following these studies, several proxies or directives for spatial resilience have been identified: diversity, redundancy, proximity, intensity, connectivity and modularity (Table 5.1).

Institutional resilience refers to the institutional ability to attain strength from the natural disasters and risks associated with the environment where humans meet with space. Institutions are an essential element of the operating environment that populate them (Cruz et al., 2015: 2). Actors within the institutions must be able to cope with extreme operating environments that are

Table 5.1 Directives of spatial resilience for public space

Directives	Description
Diversity	The space distribution of different elements. Diversity helps to build complexity and redundancy by increasing the opportunities for interactions and providing more options.
Proximity	The combined effect of good connectivity and high diversity improves proximity to opportunities and options. Good proximity reduces the average distance between places and increases efficiency.
Intensity	Intensity can be described as the quantity or concentration of something within an area at a given scale. Increased intensity of elements within an area improves access to opportunities and improves efficiency, and more options are available closer.
Connectivity	The ease of moment, which is determined by the type, quality and configuration of the movement network. Good connectivity makes movement easier while also providing more route options.

Source: Landman and Nel (2021).

unpredictable—for example, natural disasters, social upheaval, terrorist attacks and other threats.

Public spaces are the communal assets that bring together people from all walks of life and create a sociable environment that strengthens the relationship between human beings and the environment. Informal trading in the streets forms a part of the informal economy, which is an umbrella of all informal activities that take place in towns and cities. According to Troung (2018: 4), the goal of the informal sector is to secure the survival of the individual and his/her family, while that of the formal business is to maximise profits. In many developing countries, the informal economy contributes significantly to employment (Willemse, 2011: 7). However, street traders are often affected by the absence of social protection and protection against human rights violations and other social rights (Gaspar Garcia Centre, 2014: 9). This can be exacerbated by intolerance and over-regulation. The following discussion narrates a story of the informal traders in Warwick Junction to highlight these tensions.

Study area and background

The chapter draws on findings from the Victoria Street market at Warwick Junction in Durban located in the eThekwini Metropolitan Municipality. The municipality forms part of the 11 districts in the province of KwaZulu-Natal and is located on the east coast of South Africa. The municipal area comprises approximately 2,297 km² and is home to some three and a half million people. Increased migration from rural areas contributes to population growth. The eThekwini Municipal Region is the economic powerhouse of KwaZulu-Natal and also makes a significant contribution to the South African economy. It is a vital link between the regional economies of Pietermaritzburg (and onward to Gauteng) and Richards Bay towards the north. EThekwini ranks as the second-largest economic centre in the country and is the most significant industrial region (eThekwini IDP, 2018/2019).

Warwick Junction is one of the major transport hubs in the city characterised by great diversity, both in terms of people and goods. It is spatially located in the inner city with well-designed infrastructure, including transportation routes that create convenience and accessibility. This junction is well-known as a marketplace where street trading is a vital business, offering employment to thousands of people. It is estimated that between 6,000 and 8,000 of Durban's informal traders work in Warwick Junction (Alfers et al., 2016).

In 1994, South Africans elected their first democratic government on the basis of improving the lives of the poor and desire for change. The *Warwick Junction Renewal Project* then began during the post-apartheid period (the late 1990s) to provide an improved level of service to the commuters and workers who pass through daily. The project invested over 40 million Rands (approximately EUR 2 million at the 2020 exchange rate) in the area to address issues of neglect, high levels of crime and disorder. This resulted in the development of nine new markets, including the Bovine Head Cooks Market, Early Morning and Victoria Street Markets, Brook Street Market, Traditional Healers Market, Mielie Cooks Market and Bead Market (Alfers et al., 2016: 394).

Although there are many different markets in Warwick Junction, this chapter focuses on the Victoria Street market and its street traders and draws from research conducted in 2017 and 2020 for a master's and PhD study. Victoria Street (indicated with a dotted line in Figure 5.1) is located in the centre of Warwick Junction and offers access from the Berea station into the rest of the precinct on the opposite side of the National Road. It is therefore conveniently located for commuters coming from the station and moving into the rest of the precinct.

Various methods were utilised to understand the nature, use and perceptions of users in the streets. These methods included spatial analysis, site observations and semi-structured interviews. The spatial analysis was aimed to determine the nature of the physical form, including where the traders operated. The purpose of the site observations was to determine the function of the space—for example, transactions between the traders and buyers. The semi-structured interviews set out to determine the perceptions of users about the use of the spaces, including those of the traders. At least 15 respondents were interviewed in Victoria Street and the adjacent market regarding what encouraged and discouraged the use of the street. The municipal officials were also interviewed to get an insight into how they respond to street trading in the city. The aim was to discover how traders conduct their business and how the buyers and other users feel about street trading. The researcher approached a variety of users who were willing to participate and allowed for variation in age and gender.

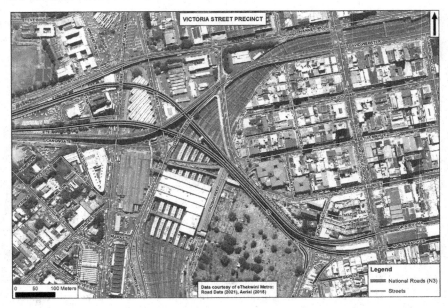

Figure 5.1 Victoria Street Precinct in the eThekwini Municipality.

Source: eThekwini Municipality, prepared for the authors by Dr Z. A. Botes.

Trading spaces and traders in Warwick Junction

Entrepreneurship

As mentioned before, Warwick Junction is predominantly an informal economic hub in the City of eThekwini that accommodates over two million people a day for different purposes. The informal nature refers to the nature of the practices and the types of trading stalls. The council recognises the traders as informal traders. However, the activities that take place in this space reflect that there are different levels of informality. This includes organised trading that takes place within the larger market spaces provided by the council, random trading that takes place anywhere in the streets outside on the pavements and, finally, the mobile traders who walk around in the streets and sell without permits.

During the interviews, the traders reflected that they have reverted to street trading as a form of entrepreneurship to make a living as formal opportunities are not available due to high levels of unemployment in the city. However, the research revealed that not all street traders are poor. Some traders were more comfortable being self-employed rather than working for an employer. It was also discovered that street traders prefer informal trading as their profit is not taxed, enabling them to pay the school fees of their dependents without any challenges. Another benefit was that while they can make sufficient money to support a family, their job is flexible. During the interviews and observations, the researcher discovered that the time spent by the traders differed. While some spent five to eight hours in their trading stations, others spent 11 to 12 hours on the street.

Many traders have engaged in trading activities for a long time. Mr N.P. indicated that he had spent over 20 years trading in the street. He relocated from his place of origin, uMtshezi to Durban in search of job opportunities. Yet, with high levels of education and competition in the city, he struggled to access employment in the formal sector. He then opted for street trading as going back home was not an option—he had children and a wife to support. N.P reflected that his decision was good as he has provided for his family over the years, paying for two children who completed university and three more finalising their secondary education. The traders in Victoria Street argued that this market is easily accessible and convenient to buy goods as it is close to various transportation modes and other markets, drawing many people. Consequently, their goods sell faster than in other areas within the city. Trading at Victoria Street thus offers sustainable livelihood opportunities.

Changing political agendas and planning visions

Pressured by changing political agendas in 1994, the City of eThekwini prioritised the refurbishment of public spaces to create a conducive environment for trading. The urban renewal projects ranged from supporting small entrepreneurs like traditional herbalists, healers and food suppliers to the provision of physical infrastructure, including bridges, roof extensions and sanitation facilities located at and under the freeway system adjacent to the rail station.

The project planning and implementation involved the participation of many stakeholders, ranging from the planners expressing their vision of a clean and safe area to the informal traders stressing the need for places to trade. Fortunately, the final idea surpassed the original attempts to suppress the informal traders and develop a pristine, formal trading area (www.pps.org/places/warwick-junction). The relationship between the traders in Warwick Junction and the local government shifted over the years, from being very antagonistic to traders during the apartheid era to being more tolerant and open to experimentation with more participatory forms of governance (Alfers et al., 2016: 394). Warwick Junction was upgraded to provide healthy, safe and secure spaces for both users and traders. Crime levels were high in the 1990s, resulting in many 'Crime Prevention through Environmental Design' principles being applied in the upgrading, including mixed-use areas around the market, design for passive surveillance (including a 'see-through' pedestrian bridge to the station), well-designed pedestrian routes, appropriate street lighting, 24-hour use in some areas and safe overnight sleeping facilities for some of the traders (Landman, 2009).

The municipality provided covered trading spaces to protect the traders from the rain and sun, which are rented for R300 to R600 per month. Over the years, the municipality ensured accessibility to basic infrastructure, such as water and sanitation, to provide a healthy environment for its users. Durban's Department of Informal Trade and Small Business Opportunities secured funds for the development of infrastructure, indicating an acknowledgement of street trading. The covered space consists of various activities that are taking place throughout the day, with the market as an anchor activity. It includes both formal businesses, such as KFC and Cambridge Foods, and informal traders in the same area. A trader, P.P. stated that in the past, the municipality allocated stalls to various traders according to the types of goods they sell. This kind of uniformity was a priority for the municipality. The space is adjacent to taxi ranks, a bus rank and a train station that provides connectivity and access to public transport. Consequently, the traders serve many commuters. According to traders, there is no specific protocol to guide how they should organise themselves in the covered market space. They have allowed the space to flow smoothly with goods of the same type in each section of the market.

However, the covered market stalls within Victoria Street are not sufficient to cater for all traders within this public space. Barbershops, laundry services and taxi rank kitchens are not provided any spaces to trade in the covered markets, and they locate themselves on the pavements to do their businesses (Figure 5.2). The traders in the street build their structures and arrange themselves to create uniformity according to the types of products they sell. Yet, there is still a challenge in terms of the nature of the space to accommodate street trading effectively in the area. Mr Okeke, a street trader working on the pavement in Victoria Street, stated that there is generally a shortage of infrastructure earmarked for street trading in the city. As a result, he struggled to get a decent space in the covered market area to sell and store his craftworks.

External conditions have also affected the activities in marketplaces, as was evident in Warwick Junction during the COVID-19 lockdown. During the hard lockdown, data were collected to ascertain the impact of the changed trading conditions in the area. J. J., who is a street trader in Warwick Junction remarked,

Figure 5.2 Traders selling food on the street.
Source: authors, 2021.

[I]n the past 23 years since I have engaged in street trading.... [I was] able to support my family. However, this COVID-19 pandemic has ... [brought a] halt to our livelihoods as the goods that we are selling are no longer in demand and one had to adapt to the changing needs.

It was observed that this period forced the majority of traders to adapt to changing demands by altering their stock to accommodate the current trends of the pandemic. It also affected the amount of profit, which makes it a challenge to sustain a living. The lockdown was, however, not entirely negative. The number of people who interact in Warwick Junction per day has slightly decreased due to government regulations, and this allows the free flow of people as opposed to the usual congestion and overcrowding. The new COVID-19-related regulations ensuring greater social distancing have created more order. They, however, limit the number of goods and services that are exchanged in the marketplace.

Regulation

Within Warwick Junction, there is street trading that is both permissible and not, depending on whether traders have a licence to trade. According to Durban's Informal Economy Policy (2001: 1), during the apartheid era, there were restricting rules that controlled street trading and the establishment of markets. However, in the 1990s, rapid deregulation and the transition in local government led to a changing policy environment. The City of eThekwini developed a policy aligned to the South African Constitution and the Bill of Rights. The aim is to promote economic development. The informal economy in this city is grounded in the overall strategic policy for economic development set out by Durban's local government, including initiatives such as revitalisation strategy, Safer Cities Projects, transport planning, SMMEs and area-based management (Durban's Informal Economy Policy, 2001: 2). The policy, therefore, offers a mechanism to ease out some of the tensions and to create more room for a participatory approach to enable opportunities for sustainable livelihoods.

To sell goods, street traders need licences. Consequently, most street traders, who were interviewed, have permits to sell on the street. The permits have all the particulars of the holder, such as name, a unique number and the rules that should be adhered to by the owner at the back of the permit. The traders stated that they renew their permits every six months. However, according to the municipal official, many of these permits are obtained to secure a trading space. Most of them do not pay monthly as stipulated at the back of the permit, and due to a lack of human capacity in the local authorities and the large number of traders, it remains a challenge to identify those who default on their payments. One of the respondents, L.J., mentioned that when she started selling on the street during the apartheid era (pre-1994), the police and their dogs always chased her. She was not allowed to sell on the street without a permit. However, in 1991, she applied for a permit, which was granted. She has been trading in Victoria Street ever since. During the interviews, it was also noted that some of the traders do not have permits, including the mobile traders walking in the streets and selling goods to people in their cars. In terms of enforcement, the metropolitan police in eThekwini randomly verify the licenses of traders around the city to ensure that they comply with regulations of the municipal by-laws. While it does not fully alleviate the challenge of identifying missing payments, it assists in making law enforcement visible daily.

The social, spatial and institutional resilience of market-making in Warwick Junction

The research findings have demonstrated the adaptive capacity of the traders, trading spaces and planners in Warwick Junction. Before the COVID-19 pandemic, the traders adapted in several ways. They organised themselves according to the types of goods they sell to strengthen their networks and create a logical and legible system of orientation in the public space. This made it easier for potential buyers to know where to find which products. They also demonstrated adaptive capacity by learning from the past and adjusting their practices—for example, through acquiring a licence or moving their location, as in the case of the mobile traders. Although it is not extraordinary for entrepreneurs to adapt to customer demand, it can be difficult for such a venture to be flexible and implement these changes. Obtaining the permits was important to avoid harassment and/or prosecution. There was also an example of the transformative capacity to overcome adversity, either personally, to cope with poverty or unemployment, or collectively, to cope with COVID-19. The pandemic and regulation changes resulted in new considerations where traders had to adapt to the changing trends and demands to be responsive to what the potential buyers needed. In this way, they created new offerings to re-establish the networks and facilitate the exchange of goods to foster both the traders and their buyers' well-being. This is reminiscent of social resilience.

The urban form and function of the public space in Warwick Junction also changed and adapted in various ways; firstly because of the urban upgrading projects but secondly due to continuous adaptation by the traders. These changes increased the diversity, redundancy, intensity and access present and hence the spatial resilience of the area. The variety of marketplaces offers a diversity of goods to buyers nearby. Redundancy is strengthened through the presence of more than one trader that sells similar products in the area, which enhances the likelihood of accessing a product should one fail to trade on a specific day. The intensity or concentration of people in the area ensures a lucrative market. For the buyers, the accessibility of the products at a reasonable price makes it more convenient to support the traders. Increased connectivity also enhances accessibility to various products nearby and a variety of paths to deal with issues such as overcrowding or, in the case of COVID-19, to accommodate social distancing. In this way, the nature and use of the public space in Warwick Junction allow the sellers and buyers to adapt in various ways to accommodate beneficial exchange for both parties.

During the urban upgrading project, the planners also had to adapt their position regarding informal trading and reconsider the nature of the project to accommodate the traders, which is an indication of institutional resilience. By developing a policy on informal trading and the provision of covered trading spaces, they opened up more opportunities for the creation of sustainable livelihoods, addressing key challenges of poverty and unemployment. Consequently, "Durban is seen as a success story in terms of what has been done for street traders" (Skinner, 1999: 16).

Resilience thinking, therefore, offers a mechanism to consider the relationships between the traders, trading spaces and planners and plan for and with change and uncertainty. Given this, planners need to become aware of the merits of resilience in practice and start implementing resilience rather than just highlighting its value (Coaffee & Lee, 2016: 263). For a long time, resilience has been considered in terms of natural disasters or climate change (for example, Resilience Alliance, 2010; Reed et al., 2013; Johnson & Blackburn, 2014) or as a replacement for sustainable development in policies or practices (Tshwane 2055 Strategy). This has often led to conceptual confusion, especially related to the notions of sustainability, resilience, vulnerability and adaptation (Meerow & Newell, 2016). Yet, resilience thinking allows one to find the best ways to engage with complex socio-ecological systems to achieve sustainability goals (Peres et al., 2017). Given this, sustainability and resilience have a complementary relationship.

Together, these theories offer a way to think about and restructure urban systems towards transformative goals. Resilience thinking offers a means to redefine sustainability as a normative position based on the aim to restore connections in the living system by healing and regenerating socio-ecological systems (Peres et al., 2017: 692). Therefore, an understanding of the social and spatial resilience of the traders and the trading spaces in Warwick Junction offers a mechanism to planners to redefine the value of informal trading (and its economic potential) and the role that public space plays in enabling trading practices through adaptive planning and management practices. This, in turn, provides a platform to develop institutional resilience. A more flexible approach can assist to address tensions between concerns for safety and means to productive living, both critical to sustainability. McLean et al. (2016: 7) notes that an understanding of how individuals and communities can successfully adapt to rapid and crisis-driven change is increasingly recognised as important both in terms of government policy and management response.

Conclusion

This chapter discussed the relationships between the traders, trading spaces and planners in Warwick Junction in eThekwini, one of the country's largest and busiest modal interchanges. It indicated that the location and nature of the area—and Victoria Street in particular—allowed opportunities for entrepreneurship and the establishment of sustainable livelihoods. The presence of many commuters created a viable market, where traders can operate from different types of spaces to address the demand. This allowed for the presence of various types of informality to coexist, ranging from those trading from covered market spaces to makeshift structures on the pavements and mobile trading in the street. Changing political agendas after 1994 and new planning visions paved the way for the incorporation of informal trading as part of the City of eThekwini's plans. Yet the level of adherence to regulation differs concerning formal permits and their payment, further highlighting the various levels of informality present in the marketplace.

The chapter centred on adaptable market-making in Warwick Junction. This is facilitated by the adaptive capacity of the socio-material relations in Warwick Junction, as evident through interactions between the traders, the trading spaces and the planners from the local authority. The traders saw the opportunity to make a living through street trading by applying for a space in the covered market area or trading in the street. Through their adaptive capacity, they focused on self-organisation to maintain a form of livelihood in the face of high levels of poverty and unemployment, which reflect a level of social resilience. Through interventions by both the traders and the planners, the public spaces could be adapted to increased diversity, proximity, intensity and connectivity—all key elements of spatial resilience (Table 5.1). Finally, the changes in space and regulations demonstrated a more accommodating approach from the local authority, reflecting institutional resilience. Yet, there is a need to recognise different types and levels of informality in planning policies and for regulation to accommodate a wide variety of informal traders operating in public space.

Resilience thinking, therefore, provides a useful way to understand the socio-material relations between the traders, the trading spaces and the local authority and more particularly, the planners. Understanding the inter-relationship between various dimensions of resilience, in this case, the socio-spatial and institutional resilience of the key role players in space offered an opportunity to move beyond the tensions between a safe and orderly versus an efficient and accessible place for trading. It allows us to unpack and understand the complex socio-spatial systems and reconsider the role of street trading in sustainable cities.

References

Alfers, L., Xulu, P. & Dobson, R. (2016), Promoting workplace health and safety in urban public space: Reflections from Durban, South Africa, *Environment & Urbanisation*, 28(2), 391–404.

Blomley, N. (2010). *Rights of Passage: Sidewalks and the Regulation of Public Flow*. Routledge.

Chen, M. (2020), COVID-19, Cities and urban informal workers: India in comparative perspective. *The Indian Journal of Labour Economics*, 63(1), 541–546.

Coaffee, J. & Lee, P. (2016), *Urban Resilience: Planning for Risk, Crisis and Uncertainty*. London: Palgrave.

Cruz, L.B., Delgado, N.A., Leca, B., and Gond J.P. (2015). Institutional resilience in extreme operating environments: The role of institutional work. *Business & Society*, 55(7), 970–1016.

Durban's Informal Economy Policy (2001), Prepared by the eThekwini Unicity Municipality.

EThekwini Metropolitan Municipality Integrated Development Plan (2018/2019), Review, Prepared by the eThekwini Municipality.

EThekwini Metropolitan Municipality Spatial Development Framework (2018/2019), Review, Prepared by the eThekwini Municipality.

Feliciotti, A., Romice, O. & Porta, S. (2017), Urban regeneration, masterplans and resilience: The case of Gorbals, Glasgow. *Urban Morphology*, 21(1), 61–79.

Gaspar Garcia Centre, (2014), *Street vendors and the right to city*. Accessed online at https://www.wiego.org/sites/default/files/publications/files/Gaspar-Garcia-Centre-Street-Vendors-Right-City.pdf

Hou, J. (2020), Guerrilla urbanism: Urban design and the practices of resistance. *Urban Design International*, 25, 117–125.

Johnson, C. & Blackburn, S. (2014), Advocacy for urban resilience: UNISDR's making cities resilient campaign. *Environment & Urbanisation*, 26(1), 29–52.

Keck, M. & Sakdapolrak, P. (2013), What is social resilience? Lessons learned and ways forward. *Erdkunde*, 5–19.

Landman, K. (2009), Boundaries, bars and barricades: Reconsidering two approaches to crime prevention in the built environment. *Journal of Architectural and Planning Research*, 26(3), 213–227.

Landman, K. (2019), *Evolving Public Space in South Africa: Towards Regenerative Space in the Post-Apartheid City*. London: Routledge.

Landman, K. & Nel, D.. (2021), Changing public spaces and urban resilience in the city of Tshwane, South Africa. *Journal of Urbanism*. https://doi.org/10.1080/17549175.2021.1936600

Lyndon, M. & Garcia, A. (2015), *Tactical Urbanism: Short-Term Action for Long-Term Change*. Washington, DC: Island Press.

Madanipour, A. (2017), *Cities in Time: Temporary Urbanism and the Future of the City*. London: Bloomsbury.

Masnavi, M.R., Gharai, F. & Hajibandeh, M. (2019), Exploring urban resilience thinking for its application in urban planning: A review of literature. *International Journal of Environmental Science and Technology*, 16, 567–582.

McLean, M., Cuetos-Bueno, J., Nedlic, O., Luckymiss, M. & Houk, P. (2016), Local stressors, resilience, and shifting baselines on coral reefs. *PloS One*, 11(11), e0166319.

Meerow, S. & Newell, J.P. (2016), Urban resilience for whom, what, when, where, and why? *Urban Geography*, 40(3), 309–329.

Meerow, S., Newell, J.P. & Stults, M. (2016), Defining urban resilience: A review. *Landscape and Urban Planning*, 147, 38–49.

Nel, D. & Landman, K. (2015), Gating in South Africa: A gated community is a tree; a city is not. In: Bagaeen, S. & Uduku, O. (Eds.), *Beyond Gated Communities* (pp. 203–226). London: Routledge.

Peres, E., du Plessis, C. & Landman, K. (2017), Unpacking a sustainable and resilient future for Tshwane. *Procedia Engineering*, 198, 690–698.

Reed, S.O., Friend, R. Toan, V.C., Thinphanga, P. Sutarto, R. & Singh, D. (2013), "Shared learning" for building urban climate resilience: Experiences from Asian cities. *Environment & Urbanisation*, 25(2), 393–412.

Resilience Alliance (2010), *Assessing Resilience in Socio-Ecological Systems: Workbook for Practitioners*.

Salat, S. & Bourdic, L. (2012), Systemic resilience of complex urban systems. *TeMA-Trimestrale Del Laboratorio Territorio Mobilità e Ambiente-TeMALab*, 5(2), 55–68.

Skinner, C. (1999). Local government in transition – a gendered analysis of trends in urban policy and practice regarding street trading in five south african cities. *CSDS Research Report No* (18).

Smith, N. & Walters, P. (2018), Desire lines and defensive architecture in modern urban environments, *Urban Studies*, 55(13), 2980–2995.

Troung, V.D. (2018), Tourism, poverty alleviation, and informal economy: The street vendors of Hanoi, Vietnam. *Tourism Recreation Research*, 43(1), 52–67.

Walker, B.H. & Salt, D. (2006), *Resilience Thinking: Sustaining Ecosystems and People in a Changing World*. Washington, DC: Island Press.

Willemse, L. (2011), Opportunities and constraints facing informal street traders: Evidence from four South African Cities. *Town & Regional Planning*, 59, 7–15.

6 La Boqueria, "the mirror of what Barcelona represents"

An analysis of public policy and the commodification of food markets

Maria Lindmäe and Marco Madella

Introduction

In March 2020, one of Spain's biggest newspapers wishfully announced that "the flight of tourists" was turning "La Boqueria into a neighbourhood market" again (Benvenuty, 2020). COVID-19 and the international mobility restrictions had led Barcelona to an anomalous situation where tourist flocks were no longer seen in the hallways of the popular market, nor on the streets surrounding it. The article was published on the first day of what was to become a 90-day lockdown, yet it formed part of a longer debate about the need to decrease the city's dependence on tourism. The negative effects of tourism in Barcelona have been discussed in recent years in academic publications, (Blanco-Romero et al., 2018; Cócola-Gant, 2018), in the local media and in informal expressions of opinion, such as street corner graffiti that shout "Tourists go home!"

In a city of 1.6 million inhabitants that receives around 30 million visitors a year (Ajuntament de Barcelona, 2019), food markets are also subject to the pressure of tourism, forcing public market managements to re-examine their commercial offerings (IMMB, 2012, 2015) and adapt them to the expectations and consumption patterns of the 'flying customers' (Varman & Costa, 2009). This is due to markets' reputation as places of 'localness' (Smithers & Joseph, 2010) where visitors may gaze into the purported backstage of local lives (Mac-Cannell, 1973). Food markets are ever more prone to replicate the principles of the experience economy and "intentionally use their services as the stage, and goods as props, to engage individual customers in a way that creates a memorable event" (Pine & Gilmore, 1998: 98). Market managements increasingly push to diversify their services by spectacularising food consumption (Di Matteo, 2019) through cooking lessons, tapas nights and food tasting events. Furthermore, the rise of food neophilia (Dimitrovski & Crespi-Vallbona, 2017; Medina, 2018) has nourished the growth of gastronomic tourism (World Tourism Organization, 2012), making food markets attractive for both local and foreign upscale visitors.

Even though tourist spending can be a resource for covering public markets' maintenance and renovation costs, the excessive dependence on foreign visitors can have negative effects. Visitors with higher purchasing power can contribute to retail gentrification, characterised by the increase in commercial rents that pushes traders to either elevate the price of their products, change them for ones

DOI: 10.4324/9781003197058-6

that allow bigger profit or, more drastically, change the location of their businesses (González & Dawson, 2015). This leads to the displacement of habitual customers and traders, many of whom are elderly and/or from lower-income groups (González & Waley, 2013). Studies also show that tourism-related gentrification (Cócola-Gant, 2018) has several impacts on people's health, including anxiety, stress and respiratory problems (Sánchez-Ledesma et al., 2020). Intensified uses of public spaces such as the market hallways, which become congested by tourist groups, also impact the mobility of the residents and, as such, their life quality.

The accumulation of such impacts on the residents' everyday practices has triggered the need to reduce tourism in order to recuperate the 'traditional essence' of places like La Boqueria so that the residents of the city could feel like it is 'theirs again', as a recent public document stated (Ajuntament de Barcelona, 2017: 4). Neighbourhood associations and unions have been especially active in forcing local governments to take action and provide measures against the negative impact of tourism on housing, proximity trade and public space (Nadeu, 2020; O'Sullivan, 2020). These developments are taking place in public markets, especially in the aftermath of the COVID-19 crisis that temporarily swept away mass tourism from many city destinations, leaving the formerly popular gastronomic sites in need of reconsidering their strategic plans and policies. Some scholars have already argued that because of the damages of COVID-19 on public space, we are now in a moment of a paradigm shift (Honey-Rosés et al., 2020), which is an opportunity to embark on radically new and bold projects. Following this idea, in this chapter, we will revise the market management policy and regulation to study what has led La Boqueria to fail as a space of collective interests to conclude the chapter with suggestions of a market policy based on proximity.

Methodology

The chapter is based on an analysis of the policy documents published by the city council and the Institute of Municipal Markets of Barcelona (IMMB) over the last 20 years. We focused on the market management strategies promoted by the IMMB, paying special attention to the ones mentioning La Boqueria. In addition, we studied the current market regulation and held semi-structured interviews with eight traders, the manager of the market and a representative of the IMMB. The interviews with traders were focused on their perception of changes that have taken place in the market in the past two decades, and how these changes influence their relationship with clients and other traders. We also asked them to evaluate the administration of the market, as well as their own capacity to take part in its management through the traders' association. In the interviews with the institutional stakeholders, we asked about the impact of tourism on the market and if there were any incentives to make La Boqueria a more inclusive public space for its surrounding populations. All the cited interviews and public documents have been translated from Catalan by the authors. The study forms part of the HERA-funded research project *Moving Marketplaces: Following the Everyday Production of Inclusive Public Space* that studies the mobility of traders and their role in the production of inclusive public spaces.

From ambulant stalls to the 'best market in the world'

> When I was small, my grandmother who lived next to the Barça football pitch would come down to La Boqueria once a week because she would save fifteen pesetas. Only this, what she could save, was a reason to buy in la Boqueria.
>
> (Interview, La Boqueria traders' association
> representative, April 2020)

La Boqueria, Barcelona's biggest and most famous retail market has never been simply a neighbourhood market. Its location on Les Rambles, the pedestrian street that lies right in the heart of the city, used to attract people from across Barcelona and its surrounding areas because of its accessible prices and great variety of products. This earned La Boqueria the status of being 'the market of the city' (Interview, IMMB representative, May 2020). The local perception of prominence has been reaffirmed by the International Congress of Public Markets (Crespi-Vallbona & Dimitrovski, 2016) and CNN (Goldberg, 2017), which have both declared La Boqueria to be the best market in the world.

The origins of La Boqueria as an open-air market date from the XIII century (Arnàs & Alsina, 2016), when peasants from the nearby areas installed their ambulant stalls close to Les Rambles to sell their agricultural products (see Table 6.1). The construction of the market building began in 1840 when modernist urban planning was emphasising the need for a steady public network of food supply that would find shelter under roofed buildings to improve the hygiene conditions (Guàrdia et al., 2015). The central location of many of the current municipal markets comes from the 1836 disentailment laws that freed central urban land from ecclesiastical ownership, allowing Barcelona to modernise its urban areas and incorporate new facilities, such as market buildings made of modern metal structures (Guàrdia et al., 2015). The ambulant stalls of La Boqueria, which used to occupy a good part of Les Rambles, were moved to the location of a convent that gave the initial name to the market—Saint Josep—which we can still read on its sumptuous Art Nouveau entrance. The network of public markets was extended at the turn of the century and in the 1950s and 1960s, with the aim of further improving the city's food supply under the poor socio-economic conditions of Franco's regime (Guàrdia et al., 2009).

It was not until the 1980s that the municipal markets became challenged by the changing demography, new consumption patterns and a new competitive form of food distribution introduced by the supermarkets (Hernández Cordero, 2017). In 2006, only 29% of groceries were bought in Barcelona's markets, in comparison to 70% in the 1920s (Guàrdia et al., 2015). To avoid an even bigger decline in the role of municipal markets as food suppliers, the city council established an autonomous body, the IMMB, to manage the city's markets in 1991. Since then, the markets of Barcelona have been governed in a public-private partnership: the IMMB is in charge of the administrative, commercial and functional aspects, while the traders' association of each market must

Table 6.1 The historical evolution of La Boqueria and the network of municipal markets

Period	Developments at the site of LaBoqueria
1217	First document relating that ambulant sales were taking place on the Plain of LaBoqueria, close to one of the entrance gates of the old city wall.
XIV–XVIII century	Les Rambles becomes the central point for the peasants from the outskirts of Barcelona to sell their agricultural products.
XVIII–XIX century	First concerns for the crowdedness of Les Rambles arise. The ambulant stalls are frequently moved and reallocated to surrounding areas when fairs, celebrations and/or official visits needed space in the street.
1836	Disentailment laws free land from ecclesiastical ownership, including the plot of the former convent of Sant Josep. This becomes the permanent location of La Boqueria market, sitting next to Les Rambles.
Early XX century	La Boqueria concentrates 40% of the sales of all the markets of Barcelona. First signs of overcrowding of the market. Construction of more public markets in the city.
1950s–1960s	Further extension of the network of municipal markets of Barcelona to improve the insufficient food supply that the city faced under Franco's regime.
1980s	First signs of decay of the municipal markets due to demographic changes, the appearance of supermarkets and changes in consumption habits.
1991	Creation of IMMB and the beginning of the public-private management model of the municipal markets. Architectural and structural modernisation of the markets begins.

Sources: Based on Arnàs and Alsina (2016) and Guàrdia et al. (2009).

contribute to the cleaning service, security and overall maintenance of the public facility.

Public-private partnership and the consolidation of the Barcelona model of markets

With the creation of the IMMB in 1991, a sustained and continuous policy of markets evaluation, renovation and promotion was established. Part of the new public-private partnership model was to motivate traders with long-term licences to contribute to the overall aesthetic appearance of the public food halls by investing in the renovation of their individual stalls. This way, the traders also became 'investors' with a "proactive attitude towards the revitalization and the success of the market" (Garriga Bosch & Garcia-Fuentes, 2015: 338). Pertaining and contributing to the traders' association became obligatory, which also allowed raising money for the maintenance of the markets and for their promotion. As of 2020, 29 of the 43 municipal markets have been renovated.

Through this approach, the city could theoretically maintain the markets under public administration, while increasing the share of private investment and guaranteeing the survival of public markets as food suppliers. The architectural modernisation of the market buildings has been a continuous process, and in the years after the 2008 crisis, the public investment reached its peak at 125 million euros (Ajuntament de Barcelona, 2011). In several markets, the refurbishing has been done in partnership with private supermarkets that obtained the right to open their businesses in one part of the public premises in exchange for an economic contribution to the reconstruction works. As a side effect of these changes, the overall number of stalls has reduced, with fewer traders providing their services through the public markets.

The first changes in the market management policy took place in parallel to the major urban regeneration projects preceding and following the 1992 Olympics in Barcelona. Following the success of the 'Barcelona model' of urban regeneration (Delgado, 2007), the city's municipal market policy became known as the "Barcelona model of management and restructuring of markets" (Ajuntament de Barcelona, 2009: 53). The prominence of the model triggered policy tourism between Barcelona and other Spanish and international cities, including Medellín and Gaza (Ajuntament de Barcelona, 2009). Apart from the visits paid by international delegations to learn about the best practices of Barcelona, the city led two European Union (EU)–funded projects: MedEmporion (2009–2012) and Urbact Markets (2012–2015), both of which were carried out through networks with other European cities. The international recognition of its market management policy allowed Barcelona to coordinate these networks of knowledge transfer and consolidate certain policy practices. Because of this, there are many similar approaches between the local policy documents and the ones created as part of the EU projects.

Policy guidelines towards food markets' diversification

One of the main policy guidelines underlined in the public documents is the need to diversify the uses of public markets. This is expected to be achieved through the creation of special areas for cookery demonstrations and by organising events such as gastronomic fairs to reach an audience that differs from the regular clientele (Ajuntament de Barcelona, 2004; IMMB, 2012, 2015). The installation of bars and restaurants is also pointed out as a way of attracting more visitors and making the markets more dynamic (IMMB, 2012; Ajuntament de Barcelona, 2014). In a similar vein, the Strategic Plan of the Markets of Barcelona emphasises the need of positioning markets as places that "generate unique experiences with both a commercial and social character" to provide "new initiatives that generate an added value to the experience of buying on the market" (Ajuntament de Barcelona, 2014: 91). Furthermore, the strategic plan suggests developing special programmes (such as gastronomic tours) and products (takeaway foods, gastronomic souvenirs) for markets that are in areas with a strong presence of tourism. In another document, it is suggested that "a market can be an excellent showcase for the neighbourhood, a place to find local products and the authentic spirit of the city", making it a "favourite destination for the modern form of tourism" (IMMB, 2015: 94). All of these guidelines point to one thing: the need of targeting a younger and more affluent type of clientele that disposes of time and money to enjoy the gastronomic experiences offered inside the markets.

As for La Boqueria, the different policy guidelines have been well adopted by its private management. The market has had several architectonical interventions, including the construction of two culinary classrooms. Over time, 21 stalls have been removed to widen the hallways and make space for a little plaza for public events. Numerous small stalls have been merged, giving way to bigger businesses and decreasing the number of small traders and competition. Regarding the gastronomic experiences, 10% of the businesses of La Boqueria are currently classified as bars and restaurants (Medina, 2018), and there are numerous private companies that organise tours in the market, combining the shopping experience with a cooking lesson. In addition, the current market management vision favours creating settings of 'high standing' for tasting exclusive products such as cured ham, oysters or sushi (Interview, La Boqueria traders' association representative, April 2020). With its status as 'the best market in the world', La Boqueria itself has also played an active role in the process of knowledge transfer and policy tourism in the previously mentioned EU market networks. In a similar spirit, it has been twinned with the Borough market of London ever since 2006, and since 2018, it forms part of the international alliance Magnificent Seven, which brings together some of the "most important markets in the world" (Boqueria, 2018) (Table 6.2).

Table 6.2 Development of market policy, 1991–2017

Period	Market policies concerning La Boqueria and the other municipal markets of Barcelona
1991–2011	Heavy public investment for renovating the municipal markets. Consolidation of the "Barcelona model of management and restructuring of markets".
2008	Creation of the new Municipal Ordinance of Markets that regulates the rights and obligations of the traders and constitutes the model of public-private management by forcing all traders to contribute to the traders' association.
2009–2017	Markets' orientation towards tourism; increasing offering of parallel leisure activities; installation of bars and restaurants; decreasing number and agglutination of stalls. Creation of various networks of knowledge exchange that allow policy mobility among markets around the world.
2015	Tourist groups are prohibited from entering La Boqueria during peak hours.
2017	Creation of a Government Measure for La Boqueria, prohibiting the on-site consumption of foods, the issuing of new licences for bars and restaurants and delimiting the sales of processed foods by reinforcing the availability of fresh products.

The continuous efforts of meeting the interests of foreign and more afflu-ent visitors who have become copious in Barcelona thanks to the exponential growth of tourism has borne fruit. Before the 2020 pandemic, the customer num-bers were up to 50,000 a day (Interview, La Boqueria traders' association repre-sentative, April 2020), which is equivalent to the population of the entire Raval neighbourhood where the market is located. However, the arrival of the more affluent customers contrasts with the otherwise humble Raval, which has a high proportion of immigrant population and a disposable household income per cap-ita of only 70% of the city's average. While the private side of market manage-ment has been pleased with the constantly growing revenues of La Boqueria, the public administration has recently grown concerned about the effects of over-tourism in the neighbourhood, and it has started to show some antagonism to the entrepreneurial market policies. These new dynamics have led to confrontations between the public and private actors functioning in the market.

Market regulation: Safeguarding the benefits of the citizenry?

The public administration has one fundamental tool for managing the markets: the Municipal Ordinance of Markets (Ajuntament de Barcelona, 2008). This legally binding document has the role of "making sure that the public municipal markets", such as La Boqueria, "truly offer the public service in the benefit of the citizenry" (Ajuntament de Barcelona, 2008, Article 21). The Ordinance is composed of 171 articles that regulate the functioning of the market, including the type of products that can be sold under each trading denomination.

Despite the periodic checks that public market agents perform to control the compliance of the Ordinance, some of the regulations are frequently violated by traders who seek legal loopholes to sell products not included in the Ordinance. A fruit and vegetable trader who has been working in La Boqueria for several decades told us that she now sells 'forbidden' products that are attractive to tour-ists. For her, tourism is the only source of income because "there is no neigh-bourhood in the Raval" (Interview, La Boqueria trader, June 2020), referring to the recent disappearance of residential customers as a result of tourism and retail gentrification (to which her own business contributes). The trader affirms that tourists require a different type of product, something that can be consumed immediately (e.g. fruit cut into pieces, salads, pies, juices), and that is what she offers. Nevertheless, her infractions of the market regulation have led to the issue of several fines and legal proceedings:

> They [the city council] don't want us to work for the tourist. (...) There has been a struggle for a long time. And they fine us. I'm on trial with the city council now. I have a fine of 800 euros, I have my own lawyer. (...) They fined me because I sell products that I shouldn't theoretically sell, such as pies. But you see, if 80% of it is vegetables, then I can sell them. And that is what I defend. If they see a meat pie, 80% of it is vegetables. (...) There are many other cases like this. Many pay the fine, but I didn't feel like paying it.
> (Interview, La Boqueria trader, June 2020)

Not fulfilling the market regulation is often a question of affluence (income from the stalls), as is the case of another interviewee. Situated in a very favourable location, this trader has been extending her businesses on the market over the last decades, and she now owns six different stalls in La Boqueria. She, too, has had negative experiences with the public administration and has had to pay fines of up to 3,000 euros (Interview, La Boqueria trader, June 2020). Affluence generated by the stalls allows for disobedience and for contravening the guidelines of the public administration. While the first trader had the means to hire a private lawyer to fight back against the city council, the second one paid fines that equal the monthly revenue of one stallholder with relative ease. This would not be the case for less affluent traders, for whom such economic penalisation could put the whole business under threat.

Failed intents of regaining public control over La Boqueria

The 2008 market regulation precedes many of the recommended policies, and despite being an extensive regulatory body, it struggles to adapt to the changes that have taken place at the market in the last decade. The case of the infractions is only one example of the inefficient dynamics of public control. The most common critique of the market is related to the excess of products that are elaborated especially for tourists: fruit juices and salads, pies, cured ham cones and other foods made for on-site consumption. As there has not been sufficient political interest to modify the entire municipal markets' regulation, in 2017, the local administration suggested the implementation of an exceptional Government Measure (Ajuntament de Barcelona, 2017) in La Boqueria.

The Government Measure outlined the shortcomings of the market and pointed out the changes that were necessary to "bring back the traditional market". Against many of the guidelines suggested in earlier policy documents, the Government Measure recommended putting further restrictions on the entrance of big groups, delimiting the percentage of processed products and reinforcing the sale of fresh products in the stalls. In addition, the on-site consumption of foods such as snacks was to be prohibited, together with the issuing of new licences for bars and restaurants inside the market. These modifications were to put an end to the 'drift' of La Boqueria, eliminate the products that "are not aimed at the normal audience" and recover "the type of traditional market that we promote and defend" (Interview, IMMB representative, May 2020).

While the Government Measure did represent a political will to provide solutions to the displacement of residential customers from the market, the measures were met with resistance. Due to the fear of a bigger conflict with the traders' association, the Government Measure has not been implemented so far (Interview, IMMB representative, May 2020). The traders' association has maintained a firm stance against the suggested measures that could increase the public administration's control over La Boqueria. The representative of the association admitted that, even though the market has struggled with the excess of tourism, the new regulations are in no way acceptable for the traders:

> We want a regulation that would be more open towards the current forms of sales, the forms that are functioning all across the world, not only in Barcelona but in the whole world. The city council, on the other hand, wants a regulation that is much more restrictive. (…) We are at complete crossroads in this question. (…) I think that we are at a moment where, if the Ordinance is implemented fully as it is written now, it is negative for La Boqueria, it is negative for the city and for the traders.
>
> (Interview, La Boqueria traders' association representative, April 2020)

The situation shows that the public administration is faced with the difficulty of introducing the new rules that would redirect the private businesses towards the current new visions for a more public market. This antagonism between the private and the public actors raises questions about the policies currently in use and the resilience of the present model for La Boqueria (or any public market at large).

Conclusion

Our analysis has shown that the market management policy has systematically promoted La Boqueria and other public markets of Barcelona as consumption spaces with the goal of diversifying the clientele. This, in turn, makes the markets more profitable for the private businesses as well as for the city council, which increases its income through the taxation of the traders' growing revenues. This policy has consisted of promoting the creation of diverse areas and activities that spectacularise food consumption. These opportunities seem to crystallise as a way to create bigger exchange value by intentionally using the market service as the stage and the customised products as props, to engage visitors in a memorable event for which they are willing to pay for (Pine and Gilmore, 1998). In addition, we have seen that the consumption opportunities are mostly targeted at a young and affluent local clientele as well as at tourists who want to experience a feeling of 'localness'. The over-crowdedness and the touristification of La Boqueria are thus partly the result of policy guidelines that explicitly recommend targeting high-income customers and tourists through special products and events. This supports the argument that retail and tourism gentrification is often led by public institutions (González & Waley, 2013; Hernández Cordero, 2017) and policies that guide public markets toward practices that may expel the very traders and resident customers whose interests the policies ought to serve in the first place. As a result, La Boqueria has ceased to be the 'market of the city', and it has rather become "the mirror of what Barcelona represents" (Interview, La Boqueria traders' association representative, April 2020)—a city that is incapable of restructuring its tourism-dependent economy that sustains precarious employment practices (Robinson et al., 2019).

The transnationally shaped market management policies that the City Council of Barcelona and the IMMB have followed ever since the 1990s have been central tools for implementing and justifying the increased privatisation of the markets, especially those with a high real estate value and central location such as La Boqueria. Furthermore, the consolidation of the Barcelona model for markets was boosted by the participation of the IMMB and La Boqueria in different international networks of knowledge transfer. In turn, the international recognition arising from such networks has provided an argumentative shield for the market management to defend its practices as something internationally recognised.

Despite—or because of—the international recognition, the lack of focus on the local socio-economic context has awoken extensive public criticism towards the market management and has forced the current government to show some initiative to counteract the policies of spectacularisation. Nevertheless, the public administration's inability to implement changes in La Boqueria speaks of the power that the private actors currently hold over the management of the public market. The growing antagonism between the public and private actors further complicates the creation of a future vision where the market could rebalance its target customers, have lesser dependence on tourism and cause fewer gentrifying effects on the residential customers and traders.

Through our study, we have seen that the socio-economic aspects of the surrounding populations have been of little importance at the time of defining the market policies and strategy. With the forecast of the post-COVID world being less dependent on tourism (Honey-Rosés et al., 2020), we believe that the market management policy should be reconsidered, focusing more on proximity trade and consumption. The current situation can support a shift of paradigm to turn La Boqueria into a socio-economic infrastructure of care and a space of collective interests (González, 2020), especially taking into account the already existing ecosystem of solidarity organisations in the surroundings of the market (see Xarxa de Suport Mutu del Raval; Tancada Migrante). After years of modernising public markets through costly renovations and by attracting middle- and high-class consumers, it could now be the time to reconsider the social value of markets and also for the lower-income residents. Now, more than ever, market managers and stallholders should look more attentively towards their residential surroundings to allow the market to recover its function as a place for the mingling of cross-class and multi-ethnic customers (Guàrdia et al., 2010).

Following Clarke (2012), we suggest that a public market policy should not only be negotiated between local politicians, council officers and market managers on the one hand and national or international politicians and consultants on the other. Rather, it should also be negotiated with local citizens and their representatives, groups, movements and organisations (Clarke, 2012). The involvement of local citizen groups not only as potential buyers but also as decision-makers could positively influence their social and political recognition and, at the same time, improve the market's role as an inclusive public space that represents collective interests (Paisaje Transversal, 2020). Turning markets' focus on proximity, rather than fictitious traditionality and localness, could lead the path towards more sustainable and responsible modes of consumption and provide the foundation of an alternative model of markets focused on proximity trade.

References

Ajuntament de Barcelona (2004), *Institut Municipal de Mercats de Baercelona. PAM 2004–2007.* https://bcnroc.ajuntament.barcelona.cat/jspui/bitstream/11703/96334/1/1567.pdf

Ajuntament de Barcelona (2008), *Ordenança Municipal Mercats. Text refós de l'Ordenança municipal de mercats.* https://ajuntament.barcelona.cat/mercats/sites/default/files/ordenan%C3%A7a%20municipal%20mercats%20%28text%20consolidat%29.pdf

Ajuntament de Barcelona (2009), *Mercats de Barcelona. Activitats 2009.* https://ajuntament.barcelona.cat/mercats/sites/default/files/Memoria%20IMMB%202009.pdf

Ajuntament de Barcelona (2011), *Mercats de Barcelona PAM 2008–2011.* https://ajuntament.barcelona.cat/mercats/sites/default/files/PAM%20Memoria2011.pdf

Ajuntament de Barcelona (2014), *Pla Estratègic Mercats de Barcelona 2015–2025.* https://ajuntament.barcelona.cat/mercats/sites/default/files/Llibre%20Pla%20Estrategic%20ok.pdf

Ajuntament de Barcelona (2017), Mesura de govern sobre preservació i millora del mercat municipal del Boqueria. https://ajuntament.barcelona.cat/economiatreball/sites/default/files/documents/mesuragovernperboqueriadefgener_2017.pdf_0.pdf

Ajuntament de Barcelona (2019), *Departamente d'Estadística i Difusió de Dades. Oficina Municipal de Dades.* Available at: www.bcn.cat/estadistica/angles/index.htm (Accessed: 1 July 2020).

Arnàs, G. & Alsina, M. (2016), *Mercats de Barcelona (segle XIX).* Barcelona: Albertí Editor.

Benvenuty, L. (2020), 'La huida de los turistas hace de la Boqueria un mercado de barrio', 14 March, *La Vanguardia.* [online]. Available at: https://www.lavanguardia.com/local/barcelona/20200314/474114805786/barcelona-mercado-boqueria-vecinos-barrio.html (Accessed: 15 March 2020).

Blanco-Romero, A. Blazquez Salom, M. & Canoves, G. (2018), Barcelona, housing rent bubble in a tourist city: Social responses and local policies. *Sustainability*, 10 (2043), 1–18.

Boqueria (2018), *Boqueria market becomes one of the 'Magnificent Seven'* [online]. Available at: http://www.boqueria.barcelona/boqueria-market-becomes-one-of-the-magnificent-seven-n-21-en (Accessed: 5 July 2020).

Clarke, N. (2012), Urban policy mobility, anti-politics, and histories of the transnational municipal movement. *Progress in Human Geography*, 36(1), 25–43.

Cócola-Gant, A. (2018), Tourism gentrification. In: Lees, L. & Phillips, M. (Eds.), *Handbook of Gentrification Studies* (pp. 281–293). Cheltenham: Edward Elgar Publishing.

Crespi-Vallbona, M. & Dimitrovski, D. (2016), Food markets visitors: A typology proposal. *British Food Journal*, 118 (4), 840–857.

Delgado, M. (2007), *La Ciudad Mentirosa: Fraude y Miseria del 'Modelo Barcelona'.* Madrid: Catarata.

Di Matteo, D. (2019), Gastronomic tourism innovations. In: Kumar Dixit, S. (Ed.), *The Routledge Handbook of Gastronomic Tourism.* (pp. 555–562) Abingdon/New York: Routledge.

Dimitrovski, D. & Crespi-Vallbona, M. (2017), Role of food neophilia in food market tourists' motivational construct: The case of La Boqueria in Barcelona, Spain. *Journal of Travel and Tourism Marketing*, 34(4), 475–487.

Garriga Bosch, S. & Garcia-Fuentes, J.M. (2015), The idealization of a 'Barcelona model' for markets renovation. In: *Localizing Urban Food Strategies: Farming Cities and Performing Rurality. 7th International Aesop Sustainable Food Planning Conference Proceedings, Torino, 7-9 October 2015* (pp. 336–342). Torino: Politecnico di Torino.

Goldberg, L. (2017), 10 of the world's best fresh markets, 30 November, *CNN*. [online]. Available at: https://edition.cnn.com/2012/07/17/travel/worlds-best-fresh-markets/index.html (Accessed: 22 May 2020).

González, S. (2020), Contested marketplaces: Retail spaces at the global urban margins. *Progress in Human Geography*, 44 (5), 877–897.

González, S. & Dawson, G. (2015), *Traditional markets under threat: Why it's happening and what traders and customers can do* [online]. Available at: http://tradmarketresearch.weebly.com/uploads/4/5/6/7/45677825/traditional_markets_under_threat-_full.pdf (Accessed: 3 May 2020).

González, S. & Waley, P. (2013), Traditional retail markets: The new gentrification rrontier. *Antipode*, 45 (4), 965–983.

Guàrdia, M. Fava, N. & Oyón, J.L. (2009), The compact city and public markets in Barcelona, Spain. In: *Sustainable Architecture and Urban Development* (pp. 335–349). Tripoli: Al-Fateh University.

Guàrdia, M. Fava, N. & Oyón, J.L. (2010), Retailing and proximity in a liveable city: The case of Barcelona public markets system. *REAL CORP, Proceedings/Tagungsband* (pp. 619–628). Vienna, 18–20 May.

Guàrdia, M. Fava, N. & Oyón, J.L. (2015), The Barcelona market system. In: Guàrdia, M. (Ed.), *Making Cities Through Markets Halls: Europe, 19th and 20th Centuries* (pp. 261–296). Barcelona: Museu d'Història de Barcelona.

Hernández Cordero, A. (2017), Los mercados públicos: espacios urbanos en disputa. *Iztapalapa Revista de Ciencias Sociales y Humanidades*, 83, pp. 165–186.

Honey-Rosés, J. et al. (2020), *The impact of COVID-19 on public space: A review of the emerging questions* [online]. Available at: https://doi.org/10.31219/osf.io/rf7xa (Accessed: 2 July 2020).

Institut Municipal de Mercats de Barcelona (2012), *The Markets of the Mediterranean. Management Models and Good Practices*. Barcelona: Ajuntament de Barcelona.

Institut Municipal de Mercats de Barcelona (2015) *Urbact Markets. Urban Markets: The Heart, Soul and Motor of Cities*. Barcelona: Ajuntament de Barcelona.

MacCannell, D. (1973), Staged authenticity: Arrangements of social space in tourist settings. *American Journal of Sociology*, 79 (3), 589–603.

Medina, X. (2018), Ir a comer en el mercado: Aprovisionamiento, consumo y restauración en la transformación de dos modelos de promoción de los mercados de Barcelona y Madrid. *Revista Española de Sociología*, 27 (2), 267–280.

Nadeu, B.L. (2020), Deserted Venice contemplates a future without tourist hordes after COVID-19, 19 June, *CNN*, [online]. Available at: https://edition.cnn.com/travel/article/venice-future-covid-19/index.html (Accessed: 19 June 2020).

O'Sullivan, F. (2020), Barcelona's latest affordable housing tool: Seize empty apartmentes', 16 July, *Bloomberg*, [online]. Available at: https://www.bloomberg.com/news/articles/2020-07-16/to-fill-vacant-units-barcelona-seizes-apartments (Accessed: 16 July 2020).

Paisaje Transversal (2020), *Llenar de barrio' los mercados municipales para reactivar el comercio local* [online]. Available at: https://paisajetransversal.org/2020/02/llenar-de-barrio-mercado-las-delicias-valladolid-comercio-local-reactivacion-dinamizacion-economia-revitalizacion/ (Accessed: 12 April 2020).

Pine, B.J. & Gilmore, J.H. (1998), Welcome to the experience economy. *Harvard Business Review*, July–August, 97–105.

Robinson, R.N.S., Martins, A., Solnet, D., & Baum, T. (2019), Sustaining precarity: Critically examining tourism and employment. *Journal of Sustainable Tourism*, 27(7), 1008–1025.

Sánchez-Ledesma, E., Vásquez-Vera, H., Sagarra, N., Peralta, A., Porthé, V., & Díez, È. (2020), Perceived pathways between tourism gentrification and health: A participatory photovoice study in the Gòtic neighborhood in Barcelona. *Social Science and Medicine*, 258, 1–13.

Smithers, J. & Joseph, A.J. (2010), The trouble with authenticity: Separating ideology from practice at the farmers' market. *Agricultural Human Values*, 27, 239–247.

Varman, R. & Costa, J.A. (2009), Competitive and cooperative behavior in embedded markets: Developing an institutional perspective on bazaars. *Journal of Retailing*, 85(4), 453–467.

World Tourism Organization (2012), *Tourism and Intangible Cultural Heritage*. Madrid: UNWTO.

7 The fluidity of a liminal marketplace
Souq Al-Ahad, Beirut

Christine Mady

Introduction

Marketplaces form an urban mosaic of people, practices and objects, where "multiple forms of sociality are enacted" as they allow for inclusion, the co-presence of differences and interactions that are rarely possible in shopping malls and other controlled spaces (Watson, 2009: 1577). Marketplaces' relation to their surroundings is linked to their connectivity, boundary definition and spatial organisation (Mermier & Peraldi, 2010). As places of sociality and gathering, if peripherally located, a marketplace's organisation could lead to social differentiation and segregation where "the market's centrality is paradoxically linked to its function as a liminal space" as a nexus hosting the vulnerable and underprivileged (Mermier, 2017: 19; González, 2020). In Beirut, Lebanon, this liminality is further linked to foreign workers, specifically Syrian refugees, bringing along stigmatisation and negative connotations such as danger, dirt and immorality (Barjes, 2014).

This chapter explores Souq Al-Ahad, Beirut's Sunday market, by investigating its value to various stakeholders on the micro-level and by considering its role within the surrounding context on the macro-level. The concept of liminality serves to understand this marketplace's dynamics. The concept of fluidity is employed to explain its positionality amidst the surrounding urban land contestations and transformations. I argue that on the micro-level, the marketplace has an in-between, liminal position spatially, socially and temporally. On the macro-level, it acquired a fluid state, changing its form and boundaries, but it is still recognisable as a marketplace amidst surrounding real estate dynamics. The author initially planned site visits and interviews in August 2020, yet these were partly affected by the COVID-19 lockdown and the Beirut port explosion on 4 August 2020. Under such circumstances, this work was replaced by desk-based research of references including recent interviews with developers in real estate magazines, and one personal interview with a real estate consultant. This exploration serves to project the marketplace's possible future scenarios.

Liminal space, fluid state

Liminal space is an ambiguous, in-between or threshold space of exchange, which fluctuates and transitions between different states (Sleto, 2017; McDowell &

DOI: 10.4324/9781003197058-7

Crooke, 2019). It borders between two domains without being part of either, to the extent that liminality with its social, spatial and temporal dimensions is compared to the process of refugees' mobility who transition between home and a new destination, separated by a threshold or border, without belonging here or there (Waardenburg et al., 2019: 940). The social dimension relates to the experiences and contestations among users engaging in practices, generating changes in rules and meanings for this liminal space (Murphy & McDowell, 2019; McDowell & Crooke, 2019). Spatially, "the appeal of a liminal space therefore lies in its hybrid and fluid nature" that is defined by an in-between area or zone, for example between different administrative boundaries (McDowell & Crooke, 2019: 327). Temporally, a liminal space responds to a trigger event such as contested surroundings—for example, a gentrifying context—and is reflected in different transitional periods of stagnation pauses, or interruptions of activities in that space (McDowell & Crooke, 2019).

Liminality is understood as the process of generating new meanings as part of transitioning, which results in reworking, transforming or building identities (Bhabha, 1994; Fourny, 2013). Accordingly, liminality provides a nexus for the "joint movement of meanings and reference systems defining places", which simultaneously enables the space to generate new possibilities or alternatives for interaction and makes it vulnerable due to its challenging norms (Fourny, 2013: 3–4). While liminality explains changes in the space itself, fluidity serves to explain the space's positionality within its broader context.

Law (2002) explains fluidity as changes in the form and materiality of an entity within a politically influenced spatial context. It is the quality of coping with uncertainty while making minor changes to the entity's boundaries and relations, allowing this entity to adapt and still operate within its intrinsic network. However, if abrupt changes occur, the entity's identity would be lost and transformed to an alternative one (Law, 2002: 99). The fluidity of marketplaces is understood as a bundle of transformations in their material and immaterial aspects related to the flows of people, goods and experiences, as well as transformations in their boundaries within a changing urban context (González, 2020), as in the case of a volatile real estate development area.

Marketplaces in Beirut

In the early 1800s, with the Ottoman period, Beirut's marketplaces or souqs formed a significant part of its public spaces (Davie, 2001). The souqs were divided according to the offered merchandise and centrally located in a concentration of religious and leisure activities. The location of the marketplaces allowed for encounter, sociality, interaction and exchange (Davie, 2001; Fischfisch, 2011). Under Ottoman rule, a municipal market was organised in the current Beirut Souks' location while only a few marketplaces were located peripherally, either within residential neighbourhoods or along main roads, responding to local needs (Davie, 2001; Fischfisch, 2011). During the French mandate period (1920–1940), urban planning projects reconfigured the city centre, demolishing its medieval urban fabric, including the marketplaces (Davie, 2001). With the constitution of the Republic of Lebanon in 1943, the consociational government and the free-market economy favoured private interests and entrepreneurialism. These trends were reflected through the economic activity boom with shops and marketplaces in Beirut's centre, and their expansion in the 1950s to 1960s beyond the centre, into multi-functional streets such as Hamra (Khalaf & Kongstad, 1973; Kassir, 2010; Mady & Chettiparamb, 2016; Mermier, 2017).

The civil war between 1975 and 1989 resulted in the city centre's demolition, the annihilation of public spaces that were considered dangerous and the displacement and segregation of the population into east and west Beirut (Tabet, 1996; Khalaf, 2002). The latter resulted in the emergence of a polycentric city governed by various political factions, with limited access to public spaces, including marketplaces (Khalaf, 2002). Consequently, new centralities and peripheries emerged. Marketplaces resurfaced post-war peripherally as popular places alongside more centrally located shopping streets, and malls, as is common in other cities where marketplaces are marginalised (Öz & Eder, 2012; Mermier, 2017). Among the popular marketplaces, which are set up regularly once per week, are Nabatiyeh in the south and Tripoli in the north of Lebanon (Barjes, 2014), Sabra in the western suburbs of Beirut and Souq Al-Ahad in its eastern suburbs at the edge of administrative Beirut (Figure 7.1).

In 1990, with the war's end, efforts were focused on the city centre's reconstruction as led by the real estate company Solidere. Alluding to the marketplaces' previous location, the Beirut Souks shopping centre was built and connected to its surrounding street network, similar to the historical souqs' feature, and unlike the introverted souqs such as in Aleppo or Istanbul (Gavin & Maluf, 1996; Moneo, 1998; Mady, 2017). The project architect Moneo intended Beirut Souks to serve as "a meeting place for the country's different communities" to revitalise "the familiar character of a souk while accommodating contemporary needs of shopping and retail" (Moneo, 1998: 263). This statement reflects Solidere's request to attract affluent customers and build on its motto: "Beirut an ancient city for the future", where archaeological and historical sites serve primarily to market Solidere's real estate development. Despite the pedestrian linkages, Beirut Souks is controlled and exclusive, and the underprivileged know that they should stop at its intangible limits. One indication of this exclusion surfaced during the protests in the summer of 2015 with the rise of the solid waste management crisis, a movement called 'Badna Nhaseb' (we want accountability). This movement opened a popular market in the city centre, symbolically providing access to the underprivileged, who otherwise frequent marketplaces such as Souq Al-Ahad (Mermier, 2017).

Souq Al-Ahad's transformation

Situated along the edge of the Beirut River between the municipalities of Beirut to the west and Sin El-Fil to the east, Souq Al-Ahad started after 1989 and covers an area of 9,300 square metres, with roughly 6,000 square metres available for stalls (Krijnen & Pelgrim, 2014; Mermier, 2017; Al-Jazeera, 2019; Abou Nader, 2019). It emerged as vendors gathered spontaneously on this open land, an easily

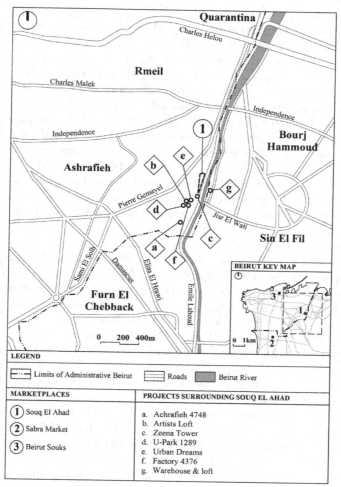

Figure 7.1 Map of marketplaces in Beirut.
Source: author.

accessible nodal location amidst three neighbourhoods: Ashrafieh in administrative Beirut, Bourj Hammoud and Sin El-Fil in the eastern suburbs (Krijnen & Pelgrim, 2014). In 2014, the Souq had around 400 stalls, supporting almost 5,000 families and attracting about 35,000 visitors on Sundays (Now Media, 2009; Krijnen & Pelgrim, 2014). While the available literature and media coverage focus on the representations and meanings attributed to this marketplace, there is the need to explore its positionality within the broader network of real estate dynamics, gentrification and political influences. The latter reflects the Souq's vulnerability, based on the threat from conflicting interests and the range of values given to its location, which are not fixed but rather fluid in space and time according to interactions among various actors (González, 2020).

Social liminality relates to the Souq's self-organisation, which shifted to a regulated one in 1996. The market was named 'Al-Souk al-Shaabi al-Kabir' (Lahoud, 2014) or the Large Popular Market. Back then two entrepreneurs, Chedid and Irani, acquired a lease from the Minister of Energy and Water to operate the market. The lease happened because of political connections entangling this marketplace "as subject to political territorialisation" (Mermier, 2017: 24). The nature of this permission remained blurred, as it could subvert laws on public properties and their use as public goods (Now Media, 2009). The entrepreneurs formed a cooperative with several Lebanese vendors, and charged fees for renting the stalls against surveillance, maintenance and cleaning services; some vendors further sublet their stalls (Krijnen & Pelgrim, 2014; Mermier, 2017). The property was initially leased for 99 years and then changed to an annual contract (Al-Akhbar, 2013; Barjes, 2014). The entrepreneurs claimed that they pay a yearly occupancy fee to the Ministry, which was a minimal fee compared to land value in the market's vicinity (Barjes, 2014).

The Souq's liminal position reflected temporally and spatially with a dispute between Sin El-Fil Municipality and the Ministry on the land's ownership and use (Now Media, 2009; L'Orient le Jour, 2011b; Krijnen & Pelgrim, 2014; Mermier, 2017). The dispute was on whether the location is east or west of the Beirut River, affecting which municipality it would fall under and whether this delimitation comes before or after the 1968 flooding mitigation measures that changed the river's boundaries (Frem, 2009). The land's ownership by Sin El-Fil Municipality was legally refuted, confirming it as public river property in Beirut, falling under the jurisdiction of the Ministry, and several decrees and legal decisions between 1924 and 1981 allowed for the leasing of such property for private use (Krijnen & Pelgrim, 2014).

However, the Municipality argued that being a public property, it should not be leased for private profit (L'Orient le Jour, 2011b). In 2007, Beirut's Court of First Instance ruled for the location's eviction and marketplace's relocation in favour of Sin El-Fil Municipality, yet the Ministry appealed, keeping the status quo (Krijnen & Pelgrim, 2014) hinting at the intervention of political influences.

Another spatio-temporal liminality was marked in 2011, with the influx of Syrian refugees fleeing the war and searching for livelihoods. For the marketplace, this meant the entrepreneurs or individual vendors sublet additional stalls to refugees, extending its premises beyond the official limits, under the bridge, to accommodate them. This extension formed a competitive divide, having a majority of Syrian vendors—some relocated from inside to pay lesser fees—selling goods at lower prices and attracting other visitors (Krijnen & Pelgrim, 2014). In 2014, Sin El-Fil Municipality described this extension as causing traffic congestion, selling all sorts of goods and giving a "bad image to the entrance" of its municipality (Mermier, 2017: 24). Its efforts to evict this extension were initially blocked by an order from Beirut's governor, until the end of 2014, curtailing the Souq's expansion and limiting congestion (Khashasho, 2014).

The dispute continued, and Sin El-Fil's mayor requested the Souq's relocation based on several complaints, reflecting its social liminality in terms of changing norms. In addition to traffic congestion, these included an absence of tax on revenue, rendering it an "illegal economic island" (L'Orient le Jour, 2011b); public and environmental health risks, noise and security issues; and the presence of Syrian vendors (Now Media, 2009; Krijnen and Pelgrim, 2014; Abou Nader, 2019). Similar to the prevailing general opinion, the mayor considered refugees a threat to local people's jobs and security, overlooking the refugees' economic disadvantages (Mermier, 2017). The Syrian vendors were discriminated against by the Lebanese ones, who considered them 'foreigners' and 'illegal', though they pay equal fees for the stalls and are subjected to regular inspections of their permits (L'Orient le Jour, 2014). This outlook was echoed in local media with reports highlighting the Souq's safety and security issues, poverty and filth, negative representations of its goods and people, disorganisation and thuggery, and some media described the marketplace as a "mark of social, economic and environmental disgrace" (Barjes, 2014; Krijnen & Pelgrim, 2014). Such portrayed images and rumours led to the stigmatisation of the vendors and merchandise (Mermier, 2017), with attributes not different from other marketplaces globally and often considered a first step towards gentrification within their contexts (González, 2020). The second-hand goods, sometimes salvaged from trash, or at other times counterfeit goods were used by some local media to characterise the Souq as a place of destitution (Al-Zaydi 2013, in: Mermier, 2017: 25). The Souq's portrayal comes in between the negative but also the positive, representing another aspect of its liminality.

Despite negative affiliations, its managers highlight the Souq's important role as a place that provides affordable goods for the economically disadvantaged and livelihood opportunities for marginalised people (Krijnen & Pelgrim, 2014; Al-Jazeera, 2019). The marketplace is truly a mosaic of merchandise with a diversity of goods including electronic, sanitary and electrical equipment, mobile phones, antiques, coins, stamps, old books, shoes, clothes and food products at competitive prices (Mermier, 2017; Al-Jazeera, 2019). Other media portray the market as attracting opposite poles on the socio-economic spectrum, with visitors ranging from foreign workers, to economically unprivileged persons but also tourists and people of different age and gender either seeking something to buy or coming for an errand (L'Orient le Jour, 2011a; Mermier, 2017). The Souq is also described as a 'flea' market and a meeting place, rich in colour and sounds, making it appealing in its own way (L'Orient le Jour, 2011a; Stoughton, 2013). Souq Al-Ahad reflected a unique entity characterised by the mix, variety and differentiation from the aspiring norms indicated by the surrounding new real estate development, and Sin El-Fil's mayor.

Surrounding transformations

Souq Al-Ahad managed to cope with uncertainty, undergoing minor changes, despite abrupt events between 2007 and 2019. The year 2007 marked the area's transformation from peripheral to central within greater Beirut, with some implications for the marketplace. Real estate development boomed post-war, resulting in Beirut's saturation with construction sites and rising land prices. This situation led developers to search for new areas, spilling over to Beirut's margins within new regions similar to Mar Mikhael, Quarantina, Badaro and the Beirut River corridor where Souq Al-Ahad is located (Boudisseau, 2011, 2018, 2020).

Similar to other marketplaces globally, occupying a border marginal site in a new central area contributed to the marketplace's vulnerability, and threatened its relocation (González, 2020: 890). In 2007, land sales in the Souq's industrial surroundings started at relatively low prices as the area was considered rough, with incomplete infrastructure, and developers described the area as a 'prime location' at the edge of Ashrafieh, a district in administrative Beirut (Boudisseau, 2009, 2011, 2018, 2020). This developers' 'playground' was favourable for constructing residential and mixed-use towers shifting from industrial uses, and projects provided some 500 apartments units, offices and commercial spaces, which signalled the area's transformation and gentrification (Now Media, 2009; Krijnen & Pelgrim, 2014; Boudisseau, 2020). Also in 2007, Sin El-Fil Municipality's unrealised proposal to evict and relocate the marketplace and convert the location into a park was blocked due to the Ministry's jurisdiction on this land (Krijnen & Pelgrim, 2014). The proposed relocation was to the area of Quarantina—Beirut's backyard east of the port—marked by foul garbage smells, where in 2000 an enclosed marketplace was set up, managed by the "Authority of Public Markets, which depends on the Council of Ministers". This marketplace was functional only in 2014 (Barjes, 2014; Mermier, 2017: 25).

I tracked real estate dynamics in 2007 to validate the claim of gentrification surrounding the marketplace. In a personal interview, Boudisseau (2020)—a real estate adviser at RAMCO—explained that land value surrounding the Souq increased between 2007 and 2014, indicating that the area was no longer cheap, and pressure on real estate was rising. Being located in Ashrafieh, I compared land prices in the Souq's vicinity to Ashrafieh (Figure 7.2).

The results confirm that the area became centralised in terms of land value in Beirut, noting that in 2020 land prices in Ashrafieh were around 1,500–2,000

Figure 7.2 Prices of land in the Souq's area compared to Ashrafieh, 2004–2016.
Source: author.

US$/BUA (Boudisseau, 2020). Becoming central required tapping on the area's features, which included high accessibility and connectivity in and around Beirut, the possibility of having a sea view similar to Beirut and the presence of landmarks such as the Beirut River or art centres. There was also a rise in local demand on smaller and cheaper apartments compared to the diaspora's earlier demand for larger ones (Boudisseau, 2011, 2018). Transformation was imminent and emerged with the new image of a business quarter, which was implemented through innovative architectural designs, creating landmarks, targeting 'sophisticated families', artists and young people, turning it into 'Beirut's Soho' (Boudisseau, 2011, 2018, 2020; Habre, 2016). In 2020, the new projects were mostly sold out. The exception, due to the financial crisis in 2019, was Zeena Tower, located directly opposite the Souq (Figure 7.1). Yet its developer optimistically promised prices to rise again once the crisis ended (Boudisseau, 2018).

In the summer of 2019, negotiations to move Souq Al-Ahad to the same location in Quarantina proposed in 2007 resurfaced, this time with an affirmative decision. Though providing affordable stall rental fees only for Lebanese vendors, parking and services, the marketplace 'building without a soul' was not a meeting place (Rozellier, 2014; El-Ghossain, 2019), and vendors refused to move. Some references indicate that a political consensus was reached to relocate the Souq, especially with the construction of residential towers whose developers were in good relations with influential politicians, specifically the owner of Zeena Tower (Abou Nader, 2019; El-Ghossain, 2019). This decision followed the Beirut governor's visit in May 2019, who built on the narrative of the Souq's legality in terms of location and practices. Shebaro (2019) explains that a consensus to relocate the Souq was reached among Sin El-Fil's mayor, Matn's governor—the governorate under which Sin El-Fil is—and deputies from the Beirut and Matn governorates. The Lebanese vendors would have an attractive premise, while the view that distorted the entrances to Beirut and the Matn for years would vanish. Also, the decision on the current location's destiny was left to the Ministry of Energy and Water, emphasising its legal jurisdiction.

The Minister of Energy and Water Boustany, who is from a different political party than Sin El-Fil's mayor, gave proposals other than relocation, which included setting new conditions for operating the current market or opening a tender for its management to avoid a monopoly (El-Ghossain, 2019). However, these alternatives were not examined. Vendors opposed the relocation, and with the national demonstrations in October 2019, eviction remained unimplemented (Abou Nader, 2019). During the COVID-19 pandemic, the market closed in March 2020 and then reopened with health and safety measures in June 2020 (An-Nahar, 2020; L'Orient le Jour, 2020). The health threat was not used as a reason to pressure on its relocation, given that the country's financial situation has led to the stagnation of most real estate operations.

Possible futures

With abrupt events starting with the war's end, to the real estate boom and the increase in real estate values in its surroundings, to the arrival of Syrian refugees,

Souq Al-Ahad's liminality enabled its fluidity, with incremental changes to its boundaries and interactions, supporting its marketplace identity amidst a changing context.

The social aspect of liminality manifested in its legal transformations from a freely organised marketplace to a contractually regulated one. The legal disputes related to its legitimacy, underpinned by political differences between the Ministry and Municipality, reflected temporal liminality, with a request for relocation in 2007, which the Souq resisted until 2020. The disputes were also caused by its spatial liminality at the border of two municipalities along the Beirut River. Despite the courts' role after 1990 to legally resolve disputes, political differences that started with the civil war continued to influence spatial dynamics and provided support to different real estate development interests (Sakr-Tierney, 2017). In its current state, the Souq is being used as a tool to maintain veiled private real estate interests (Krijnen & Pelgrim, 2014); it is a tool for 'political mobilisation' to generate alternative uses (González, 2020: 886). Depending on political affiliations and accountability, whether the Ministry or the two municipalities, differentiated management apparatuses for public or privately leased space would apply.

These uncertainties placed Souq Al-Ahad in a vulnerable, liminal position reflected in the formation of reworked categories of organisers and vendors, vendors and refugees, visitors coming for different purposes, flexible boundaries and the nature of its merchandise. The temporal aspect of liminality was evident in attempts to interrupt this marketplace by Sin El-Fil's mayor since 2007 and the stagnating, neglected character maintained by its managers. By 2014, real estate developers started considering the Souq as "garbage, as a plague or wasteland, undesired and worthy of marginalising and pushing towards a peripheral or neglected location such as Quarantina" (Boudisseau, 2020). In a newly established prestigious quarter, a popular marketplace, serving as an attraction for the marginalised and underprivileged, cannot exist, and relocating it would be inevitable.

The Municipality's call in 2007 to convert the location to a park or the vagueness of leaving the decision to the Ministry in 2019, raises questions on the land's publicness, whether it is exploitable or not: public domain or river property. Also, it indicates signs of gentrification and a response to various private interests, buttressed by political and planning decisions, reflecting how "real estate has an important role to play regarding the current and future of the market, in an area that was previously marginal but is developing as a centrality" (Boudisseau, 2020).

While the marketplace's liminality made it vulnerable to the real estate dynamics of gentrification, it also provides an opportunity for new interactions and building a new identity. This situation would transform the Souq from a negative to a positive externality, benefiting different actors. This is possible through tapping on its unique genre as a place for interaction among a mosaic of people and goods. The Souq has the potential to remain a melting pot of differences and a mix of people, operations and merchandise without being an island of contestation amidst surrounding changes (González, 2020). Considering or ignoring such opportunities might lead to future scenarios ranging from the marketplace's stagnation until complete closure, its relocation or its transformation within its current location. Souq Al-Ahad's fluid state is threatened yet could be saved if the Souq acquires a new marketplace role in its surrounding network of urban land transformations. Perhaps actors would concede to explore the possibilities of reorganising the marketplace similar to what the Minister had proposed.

References

Abou Nader, P. (2019), The true reason behind the persistence to move Souq Al-Ahad. [Al-sabab al-haqeeqi khalf al-issrar a'ala naql souq Al-Ahad] *El-Nashra*, 6 July, https://www.elnashra.com/news/show/1328007/%D8%A7%D9%84%D8%B3%D8%A8%D8%A8-%D8%A7%D9%84%D8%AD%D9%82%D9%8A%D9%82%D9%8A-%D8%AE%D9%84%D9%81-%D8%A7%D9%84%D8%A7%D8%B5%D8%B1%D8%A7%D8%B1-%D8%B9%D9%84%D9%89-%D9%86%D9%82%D9%84-%D8%B3%D9%88%D9%82-%D8%A7%D9%84%D8%A3%D8%AD%D8%AF

Al-Akhbar (2013), The evolution of Beirut's Souk al-Ahad. *Al-Akhbar*, 15 April, https://search.proquest.com/docview/1371378807?accountid=28281

Al-Jazeera (2019), 'Souq Al-Ahad in Beirut… all you need at low prices' [Souq Al-Ahad fee Beirut… Kul Ma tahtajuhu biAsa'r munkhaffidah]. *Al-Jazeera*, 2 July, https://www.aljazeera.net/news/lifestyle/2019/2/7/%D8%B3%D9%88%D9%82-%D8%A7%D9%84%D8%A3%D8%AD%D8%AF-%D9%81%D9%8A-%D8%A8%D9%8A%D8%B1%D9%88%D8%AA-%D9%85%D9%84%D8%A7%D8%B0-%D8%A7%D9%84%D9%81%D9%82%D8%B1%D8%A7%D8%A1

Al-Zaydi, I. (2013), Suriyun wa bu's lubnani fi suq al-ahad [Syrian and Lebanese Misery in Sunday Market]. *An-Nahar*, 22 November.

An-Nahar (2020), I'adat Fath Souq Al-Ahad… wa da'awat lil-iltizam bi-qawa'ed a-salamah [Reopening of Souq Al-Ahad…and calls for abiding by safety rules]. *An-Nahar*, 21 June, https://nahar.news/1215803

Barjes, I. (2014), Souq Al-Ahad: the poor are not 'thugs' [Souq Al-Ahad: al-fuqara' laysu 'baltajjiyah']. *Al Modon*, 22 September, https://www.almodon.com/economy/2014/9/22/%d8%b3%d9%88%d9%82-%d8%a7%d9%84%d8%a3%d8%ad%d8%af-%d8%a7%d9%84%d9%81%d9%82%d8%b1%d8%a7%d8%a1-%d9%84%d9%8a%d8%b3%d9%88%d8%a7-%d8%a8%d9%84%d8%b7%d8%ac%d9%8a%d8%a9

Bhabha, H. K. (1994), *The Location of Culture*. London, New York: Routledge.

Boudisseau, G. (2009), Achrafieh 4748. *Commerce du Levant*, 1 August, https://webcache.googleusercontent.com/search?q=cache:JGfxFAOBZyYJ:https://www.lecommercedulevant.com/article/16906-achrafieh-4748+&cd=5&hl=en&ct=clnk&gl=lb

Boudisseau, G. (2011), Aboudi Farkouh: « La Corniche du Fleuve, le nouveau Soho de Beyrouth ». *Commerce du Levant*, 29 December, https://webcache.googleusercontent.com/search?q=cache:ARGgrb0CvQsJ:https://www.lecommercedulevant.com/article/20016-aboudi-farkouh-la-corniche-du-fleuve-le-nouveau-soho-de-beyrouth-+&cd=5&hl=en&ct=clnk&gl=lb

Boudisseau, G. (2018), Antoine Koyess: « Les promoteurs ne font pas de profits ». *Commerce du Levant*, 2 February, https://webcache.googleusercontent.com/search?q=cache:h_k7Xo78aoIJ:https://www.lecommercedulevant.com/article/28084-antoine-koyess-les-promoteurs-ne-font-pas-de-profits-+&cd=1&hl=en&ct=clnk&gl=lb

Boudisseau, G. (2020), *Personal Interview with C. Mady*, 4 August, Beirut.

Davie, M. (2001), *Beyrouth 1825-1975 un siècle et demi d'urbanisme*. Beirut: Order of Engineers and Architects.

El-Ghossain, I. (2019), Souq Al-Ahad to Quarantine: no place for the poor in this city [Souq Al Had ila-l-Karantina: La makan lil-fuqara' fee hathihee al-madina]. *Al-Akhbar*, 20 July, https://al-akhbar.com/Community/273775

Fischfisch, A. (2011), *Formes urbaines et architecturales de Beyrouth: Depuis le XIXe Siècle jusqu'à nos jours*. Beirut: ALBA Académie Libanais des Beaux-Arts, Université de Balamand.

Fourny, M.C. (2013), The border as liminal space: A proposal for analyzing the emergence of a concept of the mobile border in the context of the Alps. *Journal of Alpine Research*, 101(2). doi: 10.4000/rga.2120.

Frem, S. (2009), *Nahr Beirut: Projections on an Infrastructural Landscape*. Master Thesis, Department of Architecture, Massachusetts Institute of Technology. http://hdl.handle.net/1721.1/49720

Gavin, A. & Maluf, R. (1996), *Beirut Reborn: The Restoration and Development of the Central District*. London: Academy Press.

González, S. (2020), Contested marketplaces: Retail spaces at the global urban margins. *Progress in Human Geography*, 44(5) 877–897.

Habre, O. (2016), Soho Beirut. *The Executive Magazine*, 28 September https://life.executive-magazine.com/life/design/architecture/soho-beirut

Kassir, S. (2010), *Beirut*. Berkley: University of California Press.

Khalaf, S. (2002), *Civil and Uncivil Violence in Lebanon: A History of Internationalization of Communal Conflict*. New York: Columbia University Press.

Khalaf, S. & Kongstad, P. (1973), *Hamra of Beirut: A Case of Rapid Urbanization*. Leiden: Brill.

Khashasho, I. (2014), The Sunday market: a secure place for non-Lebanese and the Ministry of Interior doesn't close it [Suq al-ahad bu'ra amniya li-ghayr al-lubnaniyin wa-l-dakhiliyya la taqfiluhu]. *An-Nahar*, 17 September.

Krijnen, M. & Pelgrim, R. (2014), *Fractured Space: The Case of Souk al-Ahad, Beirut*. Jadaliyya, 23 July, https://www.jadaliyya.com/Details/30990/Fractured-Space-The-Case-of-Souk-al-Ahad,-Beirut

L'Orient le Jour (2011a), Souk el-Ahad, une marche aux puces livre à lui- même. *L'Orient le Jour*, 2 May, https://search.proquest.com/docview/880719491?accountid=28281

L'Orient le Jour (2011b), La municipalite de Sin el-Fil, persona non grata. *L'Orient le Jour*, 2 May, https://search.proquest.com/docview/880719429?accountid=28281

L'Orient le Jour (2014), Perquisitions au marche du dimanche pres de Beyrouth, plusieurs personnes interpellees. *L'Orient le Jour*, 21 June, https://search.proquest.com/docview/1539207287?accountid=28281

L'Orient le Jour (2020), Coronavirus: Troix nouveau cas aux Liban, désormais 10 au total. *L'Orient le Jour*, 1 March, https://search.proquest.com/docview/2369161688?accountid=28281

Lahoud, C. (2014), Mashallah news and AMI tell tales of Beirut, past and present. *McClatchy - Tribune Business News*; Washington 11 April, https://search.proquest.com/docview/1515048089?accountid=28281

Law, J. (2002), Objects and spaces. *Theory, Culture & Society*, 19(5–6), 91–105.

Mady, C. (2017), Reading the dynamics of contemporary urban spaces in a turbulent era: A case from Beirut. *Built Environment*, 43(2), 256–271.

Mady, C. & Chettiparamb, A. (2016), Planning in the face of 'deep divisions': A view from Beirut, Lebanon. *Planning Theory*, 16(3), 296–317.

McDowell, S. & Crooke, E. (2019), Creating liminal spaces of collective possibility in divided societies: Building and burning the Temple. *Cultural Geographies*, 26(3), 323–339.

Mermier, F. (2017), Markets and marginality in Beirut. In: Chappatte, A., Freitag, U. & Lafi, N. (Eds.), *Understanding the City through its Margins: Pluridisciplinary Perspectives from Case Studies in Africa, Asia and the Middle East* (pp. 19–31). New York: Routledge.

Mermier, F. & Peraldi, M. (2010), Introduction. In: Mermier F. & Peraldi, M. (Eds.), *Mondes et places du marché en Méditerranée: Formes sociales et spatiales de l'échange* (pp. 7–17). Paris: Karthala/CJB/IFPO.

Moneo, R. (1998), The souks of Beirut. In: Rowe, P. & Sarkis, H. (Eds.), *Projecting Beirut: Episodes in the Construction and Reconstruction of a Modern City* (pp. 263–273). Munich: Prestel Verlag.

Murphy, J. & McDowell, S. (2019), Transitional optics: Exploring liminal spaces after conflict. *Urban Studies*, 56(12), 2499–2514.

Now Media (2009), 'Souq Al-Ahad – Second Part'. *Now. Media*, 1 September, https://now.mmedia.me/lb/ar/nowspecialar/%D8%B3%D9%88%D9%82_%D9%84%D8%A3%D8%AD%D8%AF_-_%D9%84%D8%AC%D8%B2_%D9%84%D8%AB%D9%86%D9%8A/

Öz, Ö. & Eder, M (2012), Rendering Istanbul's periodic bazaars invisible: Reflections on urban transformations and contested spaces. *International Journal of Urban and Regional Research*, 36(2), 297–314.

Rozellier, M. (2014), Reouverture du marché populaire de La Quarantaine. *L'Orient le Jour*, 30 December, https://search.proquest.com/docview/1640925256?accountid=28281

Sakr-Tierney, J. (2017), Real estate, banking and war: The constructions and reconstructions of Beirut. *Cities*, 69, 73–78.

Shebaro, A. (2019), After years of request, Souq AlAhad moves to quaranie [Ba'da sanawat mina-l-mutalabah souq Al-Ahad yantaqil ila Al-Karantina]. *An-Nahar*, 25 June, https://www.annahar.com/article/989116-%D8%A8%D8%B9%D8%AF-%D8%B3%D9%86%D9%88%D8%A7%D8%AA-%D9%85%D9%86-%D8%A7%D9%84%D9%85%D8%B7%D8%A7%D9%84%D8%A8%D8%A9-%D8%B3%D9%88%D9%82-%D8%A7%D9%84%D8%A3%D8%AD%D8%AF-%D9%8A%D9%86%D8%AA%D9%82%D9%84-%D8%A5%D9%84%D9%89-%D8%A7%D9%84%D9%83%D8%A7%D8%B1%D9%86%D8%AA%D9%8A%D9%86%D8%A7

Sleto, B. (2017), The liminality of open space and rhythms of the everyday in Jallah Town, Monrovia, Liberia. *Urban Studies*, 54(10), 2360–2375.

Stoughton, I. (2013), A guide to buying second hand in Beirut. *McClatchy - Tribune Business News*; Washington 12 October. https://search.proquest.com/docview/1441343102?accountid=28281

Tabet, J. (1996), *Al-I'maar Wal-Masslaha Al-A'amah [Reconstruction and the Public Good]*. Beirut: Dar Al-Jadid.

Waardenburg, M., Visschers, M. & Deelen, I. (2019), Sport in liminal spaces: The meaning of sport activities for refugees living in a reception centre. *International Review for the Sociology of Sports*, 54(8), 938–956.

Watson, S. (2009), The magic of the marketplace: Sociality in a neglected public space. *Urban Studies*, 46(8), 1577–1591.

8 Marketplace decline heads east

Neoliberal reform, socio-spatial sorting and patterns of decline at Sofia's public markets

Nikola A. Venkov

Introduction

There has been a boom of literature discussing the positive impact of street marketplaces on the city. Marketplaces are being studied as contributing to local economies and alleviating food access for low-income groups (Taylor et al., 2005), as aiding place-making and community development (Morales, 2009; Janssens & Sezer, 2013) and as engendering sociality and intercultural conviviality (Watson, 2009; Black, 2012). At the same time, policies of either restricting or restructuring marketplaces by urban authorities continue unabated (Öz & Eder, 2012; González & Waley, 2013; Guimarães, 2018). Marketplaces have become a prominent element of the gentrification frontier worldwide (González, 2020).

Plans to revitalise marketplaces, and the associated politics of displacement of their users, have typically come on the heels of discourses of 'decline'. The sense that urban marketplaces are in decline became prominent in cities around Western Europe in the late 20th century. Now, in the 21st century, a perception of decline is also taking hold in Eastern Europe. This is despite having a different historical relationship between the city and countryside, including the routine presence of marketplaces in the everyday life of urban residents throughout the 20th century (Blumberg, 2015).

In policy documents, "the decline of marketplaces" has been attributed to changing consumer expectations and a failure of marketplaces and local authorities "to catch up with the times" (González & Waley, 2013). González and Waley object, arguing instead that the present situation of traditional marketplaces (in the United Kingdom) is the result of neoliberal urban restructuring orchestrated by the state and designed to create a commodified city space. In this chapter, I flesh out their argument by exploring in ethnographic detail what neoliberal restructuring of marketplaces may look like. At the same time, their thesis is extended to the context of Eastern Europe. I interrogate the shifting dynamics at—and between—a central and a peripheral marketplace in Sofia, Bulgaria. These are the Women's Market, the largest marketplace of Sofia (Eneva, 2018; Venkov, 2018) and the Ancient Wall Market, a small neighbourhood market not far from the city centre (Venkov, 2015). The research adds to other works on post-socialist neoliberal reforms of small-scale vending (Polyak, 2013; Rekhviashvili, 2015).

DOI: 10.4324/9781003197058-8

This chapter explores how the Eastern European brand of neoliberal hegemony over city policy brought about *patterns of decline* for marketplaces. The latter term is employed here to distinguish socio-material re-orderings of urban space from the production and circulation of images of decline. In studies based in the region, discourse and images have been prominent in the analysis, especially discourses of shame as the driver of public-authority efforts for disciplining or removing marketplaces (Petrova 2011; Polyak, 2013; Venkov 2018). The present contribution fills a gap in this literature and focuses on the socio-material changes at marketplaces that have been interpreted as markers of 'decline'.

A claim advanced here is that to understand how patterns of decline emerge, one needs to consider how people are constrained to specific economic niches, social milieus and regions of urban space depending on the resources and social characteristics they possess. To this end, I emphasise processes of *socio-spatial sorting* of people, places and practices.

The notion of "socio-spatial sorting" is elaborated in the next theoretical section. Then the context of the marketplaces in post-socialist Sofia is introduced, and in the third section, I describe a neoliberal policy reform instituted in 2006. Two empirical sections demonstrate how this reform intensified several processes of socio-spatial sorting. Finally, a discussion of the findings section expands on the empirical analysis and a concluding section reflects on the wider significance of the notion "socio-spatial sorting".

Spatialising inequality: Socio-spatial sorting

The narrative of the decline of public markets in Western Europe is predicated on long-term trends of dropping customer numbers, decreasing revenues, increased trader turnover and stall vacancy, poor maintenance and the lack of reinvestment by public authorities (González & Waley, 2013; Guimarães, 2018). Here I argue that an important component of many of these patterns is movement across the city—customers dispersing, traders looking elsewhere for a means of livelihood, money flowing out, etc. Even the association of marketplaces with social stigma depended on the change of shopping trajectories in the city. Negative representations have stemmed from the increased spatial concentration of poor shoppers, the elderly, immigrants or illicit activities at the traditional marketplace. Social stigma could arise only because the middle classes had already stopped frequenting the public market and began to cultivate their sensibilities about food, hygiene and civility in different shopping contexts (Venkov, 2018: 249–257).

These changing patterns of use have been construed chiefly as the result of 'consumer preference', as the free choice to move in or move away from a particular space and type of retail. Even when urban studies have pointed out the vital functions marketplaces perform for disadvantaged groups (e.g. Taylor et al., 2005; Watson, 2009), they still have paid little attention to the modes of constraint and exclusion present in the dynamics that led to this association between the geography of marketplaces and the geography of social marginality in the city.

The purpose of this chapter is to connect the aforementioned patterns of movement and decline with a spatialised understanding of social inequality. I consider that movement means different things to different people; that it is differentiated and implicated with power; a constitutive part of it is fixities (Adey, 2006, p. 83). Manderscheid (2009) draws on Bourdieu to conceptualise inequalities as continuously re-produced and contested in multiple and relationally constituted social spaces. She draws on Massey, Urry and others to understand geographical space as socially and relationally constituted through material practice, movement and an imbrication with inequality. Building on Manderscheid's insights, I am interested in how such a continuously re-produced social inequality could attain a degree of spatial coherence and provide a spatial dimension to the unequal distributions of resources, activities and life chances.

I start from the notion of 'social sorting', used for understanding the reproduction of inequality along the course of people's lives. It has been applied to the self-sorting of populations into patterns of inequality due to the preferential association between people with similar professional status, levels of education, race, ethnicity, sexuality and so on (Bottero, 2007). In a different sense, the term has been used about the sorting of subjects by external operators such as the highly structured corporate economy in Western societies (Kerckhoff 1995) or the educational system (Domina et al., 2017). For example, schools execute multiple categorisations on pupils that powerfully affect their chances later in life.

The idea of 'social sorting' impacting subjects' life paths speaks to the mode of constraint and exclusion that I am seeking to uncover in the patterns of movement and change of use in the city. However, I open up the formulation to include 'self-driven' processes of categorisation in the context of relational dynamics (e.g. competition) taking place across a varied population of subjects. The theoretical notion proposed here, "socio-spatial sorting", adds also awareness to spatiality, to a continuously re-produced urban geography and the multidimensional character of inequality (Manderscheid, 2009).

Socio-spatial sorting captures processes of socio-economic and spatial reordering composed by the actions of a large number of agents placed in a relational setting of competition. Competition essentially acts as a dynamic filter that sends subjects one way or another depending on the unequal resources and varied social characteristics they are equipped with. Still, a diversity of outcomes survives because individuals draw on different, creative strategies and have different, complex backgrounds that determine if they will follow a certain path or not. There are moments when a 'pattern' of movement may be said to emerge as prevalent, although not holding absolutely. For example, one could claim that "better off consumers are leaving the open-air marketplace for the mall", although consumers of means would still be found at the marketplace. It is to such tentative patterns emerging in the messiness of the social that the notion of socio-spatial sorting aims to draw attention to.

Socio-spatial sorting lets us look into how specific spaces might end up with higher concentrations than others of certain types of users, activities, relationships. Unlike a notion that has been used in the study of residential segregation, "socio-spatial differentiation" (Li, 2019), socio-spatial sorting avoids taking inequality as an independent input variable. It sees the production of urban inequalities as simultaneously constituting and being constituted by the relational practice that shapes urban space itself.

The following sections explore some of the concrete forms socio-spatial sorting may take concerning urban marketplaces. They substantiate how the sorting of different social groups and practices into and out of the public market produces or exacerbates patterns of decline and demonstrate that marketplace decline has made inroads in Eastern Europe by the first decade of the 21st century. The discussion is informed by my long-term research engagement with the Women's Market (2010–2013) and the Ancient Wall Market (2014). It was by the early 2010s when the patterns of decline identified in this chapter reached their peak and helped the subsequent city visions for marketplace redevelopment gain the upper hand. The data I collected included observation in situ, surveys of market visitors and market traders, in-depth interviews with a wide range of stakeholders and a review of policy documents, as well as the media coverage.

Sofia's evolving marketplaces

In Eastern Europe, public markets embodied uninterrupted direct contact between the city and its rural hinterland throughout the 20th century (Blumberg, 2015). In the Southeast-European socialist states, open-air markets were seen as important institutions for providing food and other goods, and they were well cared for by the socialist state (Venkov, 2021). At the socialist market, individual peasants, as well as state cooperatives, could offer their own produce directly to urban consumers. Peasants were only required to pay a modest fee on the day in order to secure a stall as well as the provision of scales, an apron, etc.

With the retrenchment of the socialist state in the 1990s, Bulgaria's economy collapsed, but the role of marketplaces expanded. They became hubs of informal commerce and small-scale entrepreneurship (see e.g. Konstantinov, 1996). Marketplace areas, trader and visitor numbers exploded. It was only in the 2000s when large, agglomerated actors of the Western-style consumer capitalist economy began penetrating the retail sector in Bulgaria (Tasheva-Petrova, 2016). By the end of the decade, a wave of supermarket and shopping centre investment had swept over the retail landscape. More and more customers were drawn away from the marketplace. At the same time, public policy shifted from tolerating small-scale vending to actively 'cleaning up' urban space from the latter (see e.g. Petrova, 2011). In the mid-2000s, the public marketplace found itself without many allies in the city government.

Neoliberal policy

In the 1990s, city markets were re-organised into municipally owned companies whose boards of directors answered to the city council. When the boom of the informal economy began, the money flowing through the markets sharply increased. This situation got the market management (and other institutions locally, such as the police or the sanitation authorities) involved in various corrupt practices, some of which were related to the allocation of stalls and shop booths (as evidenced in many interviews I conducted). Formally, traders were still obligated to pay a fixed daily or monthly fee as in the socialist period, but now there were various additional informal arrangements involved. The market companies became widely known as sources of unaccounted for 'slush funds' for the political parties in Sofia (ACCESS-Sofia, 2005).

In this context, a newly elected city council took to reforming the management of markets in 2006 with the ostensible goal to make municipal company dealings more transparent. Yet, the main thrust of the changes was to start a neoliberal reorganisation of relations at the marketplace in keeping with the free-market dogma held dear by the city administration (as my interviews with city councillors and market officials attest). Scrapped were the social policies that survived even the 1990s, such as an obligatory quota of stalls for people with disabilities or a stall fee reduction for people living next to the market (Sofia City Council, 2018).

However, the key reform was the institution of a mechanism for competition in the allocation of trading lots. From now on, the rents for traders were to be determined by public tenders. Traders submitted a declaration in a sealed envelope with the monthly rate they promised to pay for their chosen lot and had no knowledge of what other traders offered for the same spot. This situation forced traders to pledge rents that were at the top of their earning abilities to avoid suddenly losing their business. Stalls were to be hired for a year; shops and booths for three years. Any business which worked with a horizon of operation longer than that was under great pressure around the time of new tenders. In this way, by pitting traders against each other, the city government managed to set up a state of "continuous commodification" of marketplace space (Öz & Eder, 2012: 307–308).

This shift in policy and attitude is illustrated by the following response made by the head of the municipal company managing the Ancient Wall Market. When I asked if he had any protections in place for long-established traders or for actual food growers he replied,

> We are a trading company, and we should seek maximum profit. We do not have social welfare capacities.
>
> (Middle-aged man, high-level municipal administration, 2014)

For the sake of contrast, just a few years before the reform, an avowedly right-wing mayor defended the lagging revenue from municipal public markets in starkly different terms:

> Most municipal companies are not set up to make a profit but to offer certain services. If we burden these companies with the task of bringing in revenues, they will cease to provide services for the citizens.
>
> (Media interview with Sofia's interim mayor, 2004, cited in ACCESS-Sofia, 2005: 161)

The following sections explore how the turn to extreme free-market principles impacted the markets of Sofia.

The food growers leaving

The new system of determining the rent by sealed auctions combated corruption by simply removing any surplus revenue from traders at public markets. If one did not pledge the highest amount that one could recoup by working seven days a week, ten to 12 hours a day, at a given stall or shop booth, one risked losing their business because of another trader's higher bid. This situation was a significant factor that squeezed out actual agricultural producers from the marketplaces and with them many shoppers:

> We had friends here from a village near Plovdiv. They gave up their business.... But how can that be possible?! A producer can't turn over that kind of money! The shop booths around here cost a thousand leva [roughly 500 euro], even up to two thousand! Come on then, tell me what kind of crop this guy should plant, what kind of pesticides should he use, how will he bring it over here, how will he pay a salary [for a daily vendor] and have a little something left over for himself—and on top of all that to pay 24,000 leva rent for the year?!
>
> (A middle-aged woman, a booth shopholder selling confectionery at the Women's Market, 2012)

Those who invested in a full-time presence at the market were better placed to acquire crucial informal knowledge and networks locally. Professionalising as an entrepreneur at the market allowed one to make the most of the auction system, as well as of many other opportunities. Full-time presence could be achieved by re-selling produce from the city wholesale rather than by investing one's time and effort into farming. Thus, resellers could offer higher bids than food growers who in turn felt pressured by the rising rents. The new policy started a process of sorting the market occupants towards a dominant presence of resellers.

The auction's system impacted the composition of traders at the market (authentic food growers vs. resellers; Figure 8.1), but more than that the composition of practices. The agricultural producers who persisted at the market had to switch to retailing too when their own produce was not in season. In turn, the difficulty of juggling the labour needed at one's village and at the market, and securing stock at the wholesale (every day before dawn), meant that producers still present at the market were mainly sorted into two types. A few traditional food growers managed to mobilise an extended family in a complex spatio-temporal arrangement. For example, a married daughter and her family would settle permanently in the city and work as resellers while the others would remain based in the village and, in season, would be supplying the family in Sofia with stock. The other type consisted of larger farmers who had enough resources to operate as a company and hire vendors and a delivery man

Figure 8.1 A mix of agricultural producers and retailers at the Women's Market, 2012.
Source: author.

as employees in Sofia. When the company's own crops were not in season, the Sofia employees would run the same retail business as their other colleagues at the market.

In this way, the shopping experience at the marketplace was gradually transformed from buying directly from food growers to buying from wholesale stock. There was little to distinguish the goods from those at neighbourhood fruit and veg shops. This is the first significant "pattern of decline" identified in this chapter. For most consumers, not being able to recognise a food grower at the stall made them highly distrustful of the products on sale. At that point in time, imported produce was flooding the Bulgarian market, although it was seen as being of substantially lower quality. It was industrially farmed and based on plant varieties developed for the needs of transcontinental shipment and a corporate capitalist cycle. However, at wholesale, these imported varieties competed successfully for the attention of the resellers due to lower, European Union–subsidised offering prices and extended 'shelf life'. Thus, the prevalence of retail at the marketplace meant that it gradually lost its privileged reputation as a provider of fresh local produce and varieties.

This is how the sorting of small agricultural producers out of Sofia's marketplaces motivated the departure of many formerly regular customers. Crowds at the markets were getting visibly thinner in another pattern of decline. The perception of decline was made more intense by the comparison on everyone's mind with the ideal marketplace of the socialist period and the first years after 1989. Such comparisons never took long to pop up in my interviews with Sofia's residents. Even immediate neighbours, who had once suffered the nuisances of a busy marketplace just under their windows, concurred:

> When was the best period of the market here? [*She laughs at the paradox:*] When the market was all just shacks filled with rats! But then they were coming over from the village with no less but car trailers full of fresh stock!
> (A middle-aged woman, living in a block of flats adjacent to the Ancient Wall Market, 2014)

Re-sorting the poor across the city

The combination of falling numbers of food growers, falling footfall and the treatment of markets as cash cows for the city budget brought another re-ordering of the patterns of movement and shopping, this time on the city scale. It intensified a central-periphery relation between the Women's Market (the largest marketplace of Sofia) and smaller neighbourhood marketplaces. Earlier, village producers considered few factors in choosing which market to attend, often just the convenience of reaching the respective location—as did the shoppers. However, relational dynamics between marketplaces became significant by the mid-2000s. The Women's Market central position became more pronounced as its bigger size and bustle meant larger turnovers and the ability of traders to stick to low margins despite rents rising with the new municipal policy:

> Here is where the trade is! Any place which is not the [Women's] Market doesn't have trade! With this stock of mine, if I moved 10 metres to that side [pointing at the off-street], I would be dead within a month! Because I work with a 5 to 7 percent mark-up price. Over there I would be forced to up it to 50% at the very least.
> (The trader of confectionery from an earlier quote, 2012)

Paying rents per square metre higher than those paid by international brand stores on the high street (Presa, 2015), professionalised traders at the Women's Market could still offer *half* the supermarket's prices. Traders at neighbourhood markets, on the other hand, were forced to enter a spiral of compensating falling turnovers by raising price mark-ups. As a result, across the city, residents with restricted budgets found themselves excluded from their local marketplace—in a country where 40% of the population lives at risk of material deprivation (NSI, 2014: 6). Poorer citizens from all around the city assembled at the Women's Market (Figure 8.2), sustaining between 40,000 and 70,000 visits per day in 2012–2013 (as reported by an internal study for the management company to which I was kindly given access to).

Confirming this picture, my survey at the Ancient Wall Market in 2014 revealed that even residents living directly adjacent to the market shopped at 'independent' fruit and veg shops located outside the market, while the elderly, who had much tighter budgets, made lengthy trips by public transport to the

Figure 8.2 Senior citizens queuing at the shop booths of the Women's Market, 2013.
Source: author.

Women's Market. (There were still no supermarket chains in the vicinity, but some better-off families had switched to driving to distant ones.)

The Ancient Wall Market by that time predominantly catered for the employees of nearby large institutions at lunch break. Many of the kiosks had been taken over by fast-food enterprises and tobacco and drinks shops. Students from the Architecture University would visit for a döner kebab or a pastry. Employees from the City Court of Justice would come to get an individual fruit in addition to their coffee or sandwich. Once a site packed with producers from the village (see quote in the preceding section), now the Ancient Wall Market boasted just eight fruit and veg vendors, only one of whom was an actual food grower, an elderly woman from a nearby village who supplemented her pension in this way. The relational impact on rents by higher-earning competitors (fast-food businesses or tobacco and drinks corporate chains) also generated a long stretch of unoccupied shop booths, shaping another pattern of marketplace decline—vacancy.

The evolving composition of users of the marketplace can be accompanied also by a change in the predominant patterns of interaction between shoppers and vendors. Such a "socio-spatial sorting of practices" can be gleaned from seemingly inconsequential ethnographic details. An example is the shift by which buying a single (expensive) apple had become acceptable at the Ancient Wall Market, even though in Bulgaria fruit and veg are traditionally sold by the kilogram. The Women's Market sat at the other extreme: there asking to buy less than a kilo of produce was asking for a rather rude response from the vendor. Such practices are enmeshed in complex economic, class and cultural linkages that could acquire a spatial dimension through the process of sorting.

In conclusion, through unanticipated relational dynamics emerging on the city-wide scale, the policy of public tenders contributed to the socio-spatial sorting of prices, uses, users, practices and places. Poorer citizens were excluded from their local markets and "sorted towards" shopping at the Women's Market. Places, in their turn, were being pulled into the realms of either vacancy or social stigma, both of which could be read by urbanites as 'decline'. Namely, smaller neighbourhood markets seemed in the process of dying, with quieter ambience than before, the odd customer, high prices and with the number of stallholders dropping off over the years. On the other hand, the central market solidified a reputation as a zone of the poor and the miserable, despite it being still very mixed, with over 500 traders and many types of visitors. In fact, it was the tensions of intense social and class mixing there—what I have elsewhere defined as "social multiculture" (Venkov, 2019)—that directly provoked the city's plans for regeneration and "cleansing of the area" in 2013–2014 (Venkov, 2018).

Discussion of findings

The patterns reviewed in the previous sections are all examples of *patterns of decline*—they are material re-orderings of space that turn away traders and customers while strengthening images of decline. The empirical observations of this chapter included the departure of agricultural producers who once made up the identity of Sofia's marketplaces, an emerging association between the market and poverty as a result of the spatial concentration of the poor at the chief marketplace of the city and, finally, the visibly reduced visitor crowds and even the blight of shop vacancy. At places such as the Ancient Wall Market, even the core marketplace identity of fruit and veg sales was affected by the takeover by more 'efficient' types of business.

While each of these changes was the result of complex dynamics and some might have run their course anyway, here I have traced how a single shift in city policy intensified them. The introduction of public tenders in 2006 imprinted the administration's neoliberal vision onto urban space. By seeing marketplaces only as sets of plots to be rented out at the highest opportune price, it set up an ever-present precarity for traders. It instituted what has been termed by Öz & Eder, 2012 "continuous commodification" of space. Through the resultant soaring rents, the municipality succeeded to appropriate an ever-larger portion of the incomes of the city's poorest shoppers and struggling micro-businesses.

The analysis in this chapter substantiates in ethnographic detail how state-led neoliberal encroachment on urban space produces the decline of the public marketplace, which then triggers politics of revitalisation and displacement (as in González and Waley, 2013). Patterns such as the ones identified in this chapter are typically drawn upon by public authorities to argue that traditional marketplaces do not suit the needs of the contemporary consumer anymore and should be transformed for use by different—invariably, more affluent—demographics.

Indeed, at both marketplaces explored here, following the period under discussion, the city-initiated projects for revitalisation, which marginalised its traditional users and uses further. A major redevelopment of the Women's Market into a pedestrian shopping street took place in 2013–2014. Today, periodic vinyl sales and gourmet street fests (see Petrova, 2020) are organised next to a public market greatly reduced in size. Interestingly, these events take place behind tall temporary fences that serve as a barrier against the area's ordinary users. As a means to 'revive' the Ancient Wall Market, weekly organic 'farmer's markets' with quite a different class composition of vendors and patrons was promoted over regular, rent-paying traders (see Venkov, 2015). By 2020, a single daily fruit and veg retailer survived at the site.

Conclusion

The literature on marketplaces and retail gentrification has powerfully contested the thesis that public marketplaces have lost their function in contemporary consumer society. However, it has tended to focus only briefly on the purported phase of marketplace decline, mainly as a prelude to discussions on the subsequent events of regeneration or gentrification. Often, multiple causes for the historical decline are listed in a rather cursory manner—the expansion of other forms of retail, suburbanisation, long-term disinvestment, neoliberal urban restructuring, power struggles and investor interests, to name a few (Öz & Eder, 2012; González & Waley, 2013; Polyak, 2013; Guimarães, 2018; González, 2020).

The present contribution identifies a single potential factor—a change in market management regulations—and teases out the causative connections linking it to patterns of decline at Sofia's public marketplaces. Rather than arguing against the importance of multiple interwoven factors for urban change, this approach suggests a methodology to make legible a multiplicity of interlinked relational fields.

Central to the contribution is the notion of "socio-spatial sorting". It captures the dynamics of spatial stratification of citizens into different regions of urban space owing to the unequal resources and social traits they are equipped with. Socio-spatial sorting is a heuristic designed to scrutinise how social inequality obtains spatial coherence and how urban geographies of marginality take shape. Fundamental to this perspective is the understanding of urban space as produced by the dynamic cross-dependency of many relational fields through which social agents are continuously juxtaposed—e.g. in their possession of economic, cultural and social capitals (if we use Bourdieu's terms), choice of tactics as well as life-long strategies, etc.

This chapter begins to sketch out the kinds of insight socio-spatial sorting can provide. Further work would look into the multiple factors that bring about patterns of change at urban marketplaces; it would uncover further spatial scales at which sorting processes play out. The full potential of the heuristic socio-spatial sorting would be explored in other thematic areas of urban studies. Gentrification is one obvious area of application since at its heart lie processes of socio-spatial sorting where local residents have found themselves at a disadvantage compared to urban space users with a different set of resources. Gentrification theory has been criticised for applying a pre-defined set of relational fields to contexts around the world where they may not be relevant (Bernt,

2016; Lawton 2020). Socio-spatial sorting, on the other hand, is adapted for cases where new relational scales emerge and interact with existing ones, as has been shown in this chapter. Furthermore, this perspective is especially useful for studying instances of neoliberal restructuring, which often implies the commodification of new regions of social space by the setting up of new scales of competition.

References

ACCESS-Sofia (2005), *Sofia – A Dirty Dream Story. Reports by ACCESS-Sofia Foundation for the Economic Activities in the Capital [София – мръсна приказка - Доклади на фондация "АКСЕС - София" за стопанските дейности в столицата].* Sofia: Apostrofi.

Adey, P. (2006), If mobility is everything then it is nothing: Towards a relational politics of (im)mobilities. *Mobilities*, 1(1), 75–94.

Bernt, M. (2016), Very particular, or rather universal? Gentrification through the lenses of Ghertner and López-Morales. *City*, 20(4), 637–644.

Black, R.E. (2012), *Porta Palazzo: The Anthropology of an Italian Market.* Philadelphia: University of Pennsylvania Press.

Blumberg, R. (2015), Geographies of reconnection at the marketplace. *Journal of Baltic Studies*, 46(3), 299–318.

Bottero, W. (2007), Social inequality and interaction. *Sociology Compass*, 1(2), 814–831.

Domina, T., Penner, A. & Penner, E. (2017), Categorical inequality: Schools as sorting machines. *Annual Review of Sociology*, 43, 311–330.

Eneva, S.A. (2018), Contested identities and ethnicities in the marketplace: Sofia's city centre between the East and the West of Europe. In: S. González (Ed.), *Contested Markets, Contested Cities: Gentrification and Urban Justice in Retail Spaces* (pp. 166–181). London/New York: Routledge.

González, S. (2020), Contested marketplaces: Retail spaces at the global urban margins. *Progress in Human Geography*, 44(5), 877–897.

González, S. & Waley, P. (2013), Traditional retail markets: The new gentrification frontier? *Antipode*, 45(4), 965–983.

Guimarães, P.P.C. (2018), The transformation of retail markets in Lisbon: An analysis through the lens of retail gentrification. *European Planning Studies European Planning Studies*, 26(7), 1450–1470.

Janssens, F. & Sezer, C. (2013), 'Flying markets' activating public spaces in Amsterdam. *Built Environment*, 39(2), 245–260.

Kerckhoff, A.C. (1995), Institutional arrangements and stratification processes in industrial societies. *Annual Review of Sociology*, 21(1), 323–347.

Konstantinov, Y. (1996), Patterns of reinterpretation: Trader-tourism in the Balkans (Bulgaria) as a picaresque metaphorical enactment of post-totalitarianism. *American Ethnologist*, 23(4), 762–782.

Lawton, P. (2020), Unbounding Gentrification Theory: Multidimensional Space, Networks and Relational Approaches. *Regional Studies*, 54(2), 268–279.

Li, Z. (2019), Sociospatial differentiation. In: A. M. Orum (Ed.), *The Wiley Blackwell Encyclopedia of Urban and Regional Studies.* (pp. 1920–1923) Oxford: Wiley-Blackwell.

Manderscheid, K. (2009), Integrating space and mobilities into the analysis of social inequality. *Distinktion: Scandinavian Journal of Social Theory*, 10(1), 7–27.

Morales, A. (2009), Public markets as community development tools. *Journal of Planning Education and Research*, 28, 426–440.

NSI (2014), *Poverty and Social Inclusion Indicators in 2014*. Sofia: National Statistical Institute.

Öz, O. & Eder, M. (2012), Rendering Istanbul's periodic bazaars invisible: Reflections on urban transformation and contested space. *International Journal of Urban and Regional Research*, 36(2), 297–314.

Petrova, V. (2011), Take the market out of sight! *Seminar_BG*, 2(1). https://www.seminar-bg.eu/spisanie-seminar-bg/special-issue1/item/319-take-the-market-out-of-sight.html

Petrova, V. (2020) Experiencing the spectacle of fine dining. New forms of festivity in Sofia, Bulgaria and diversion of public space. In: V. Marinescu (Ed.), *Food, Nutrition and the Media* (pp. 189–201). Cham: Springer.

Polyak, L. (2013), Exchange in the street: Rethinking open-air markets in Budapest. In: A. Madanipour, S. Knierbein & A. Degros (Eds.), *Public Space and the Challenges of Urban Transformation in Europe* (pp. 60–72). London/New York: Routledge.

Presa (2015, May 15), It is the traders who brought the rents up for themselves at the Women's Market [Търговците сами си вдигнаха наемите на Женския пазар]. *Presa*. http://presa.bg/article/archive/53023/2/0

Rekhviashvili, L. (2015), Marketization and the public-private divide: Contestations between the state and the petty traders over the access to public space in Tbilisi. *International Journal of Sociology and Social Policy*, 35(7/8), 478–496.

Sofia City Council (2018), Sofia, Ordinance for the markets on the territory of Sofia Municipality [Наредба за пазарите на територията на Столична община] (May 5, 2018). https://sofia.obshtini.bg/doc/114205

Tasheva-Petrova, M. (2016). Retail Gentrification and Urban regeneration of the City of Sofia. Retrospective and Perspective. Paper presented at the *"From CONTESTED_ CITIES to Global Urban Justice" Conference*, Madrid.

Taylor, J., Madrick, M. & Collin, S. (2005), *Trading Places: The Local Economic Impact of Street Produce and Farmers' Markets*. London: New Economics Foundation.

Venkov, N.A. (2015), *"Rimskata Stena" Marketplace – A Sociological Analysis/Policies Towards Municipal Markets* [Пазарът "Римска стена"– социологически анализ/ Политики към общинските пазари]. Sofia: Studio Architectonica/Sofia Municipality. https://tinyurl.com/2cj8bnzd

Venkov, N.A. (2018), Assembling the post-socialist marketplace: Transitions and regeneration projects at the central pazar of Sofia. *Ethno Anthropo Zoom/ЕтноАнтропоЗум*, 17, 229–279.

Venkov, N.A. (2019), Conviviality vs politics of coexistence: Going beyond the global North. *CAS Sofia Working Paper Series*, 11, 1–42.

Venkov, N.A. (2021), The "cooperative marketplace" as a point of pride and a challenge for the socialist system [Кооперативният пазар като гордост и предизвикателство за социалистическата система]. In: A. Luleva, I. Petrova, P. Petrov & Y. Yancheva (Eds.), *Bularian Socialism: Ideology, Everyday Life, Memory* [Българският социализъм: идеология, всекидневие, памет] (pp. 39–61). Sofia: Prof. Marin Drinov. https://tinyurl.com/yc8n7v53

Watson, S. (2009), The magic of the marketplace: Sociality in a neglected public space. *Urban Studies*, 46(8), 1577–1591.

9 Government's representation of Belo Horizonte's public markets

The (ir)reconcilable grammars of economic pragmatism and social justice

Patrícia Schappo

Introduction

Belo Horizonte, the capital of Minas Gerais state, Brazil, has a culture of markets. Citizens are fond of traditional trading such as street markets and crafts fairs. Yet, its enclosed public marketplaces, the focus of this study, are under threat. Institutionally assigned to the Food Security and Nutrition Secretariat (SUSAN), within the Social Services (SMASAC) department, markets no longer significantly fulfil their original role of food suppliers. The competition with more modern providers, supermarkets and cheaper 'buy-in-bulk' shops (in Brazilian context called '*sacolões*') rendered markets quite marginal within SUSAN programmes. Without local governments' support, they have been struggling against decay for the last decades, a tendency that hampers their contribution to social justice through, for instance, the provision of livelihoods.

Municipalities' understanding of markets' plural functions (e.g. Janssens & Sezer, 2013; Schappo & van Melik, 2017) shapes their support to markets. This chapter discusses Belo Horizonte's local government's understanding and representation of public markets' urban relevance and to what extent this perception connects markets to the attainment of social justice in the city. It addresses the markets' ongoing transfer to private-sector management, which happened between 2017 and early 2021. The process and conditions of the adopted tendering indicate a vision where markets and social justice aims overlap to a certain extent. I will explore these linkages, their potentials and limitations through Fainstein's *democracy, equity* and *diversity* justice triad (2014).

The chapter begins with a theoretical discussion of the parallels between social justice and marketplaces' urban functions and of the relevance of local governments' discourses and policies for attaining social justice through markets' presence in cities. The second section discusses management change in Belo Horizonte's public markets in relation to the adopted theoretical framework. Lastly, the conclusion addresses how far the tendering process and requirements intend to deliver markets aligned with social justice aspirations, highlighting the crucial role played by council managers in legitimising the adopted governance strategy and in further guaranteeing its proper enforcement.

This chapter draws on PhD research exploring marketplaces as social justice instruments for cities. The data is composed of ten interviews with city council

DOI: 10.4324/9781003197058-9

staff members, interviews with market traders and users, ethnographic fieldnotes, the recording of two public sessions and documents collected between July 2019 and February 2021.

Social justice, marketplaces and local governance

Social justice is a broad and contested concept, but some common ground between varied standpoints exist: it is context-dependent, requiring an understanding of the local specificities within which relations between individuals, groups and institutions can be assessed as just or unjust (e.g. DESA, 2006; Fainstein, 2014; Harvey, 1973). Moreover, there is an intrinsic relation with the idea of a justly arrived at *distribution* of benefits and burdens (Harvey, 1973: 98).

Despite critics like Hayek (1979) who consider social justice an empty signifier, the debate about what it entails and how to achieve it has kept evolving since the 1970s, sustained by increasing exclusionary global realities under neoliberalism. Philosophy, political science, geography and planning moved from purely material and economic considerations of inequality and distribution to an ideal concerned with recognising difference (Young, 1990) and the participation of multiple groups in policymaking. The United Nations stated its commitment to social justice in 2006, defining it as the fair and compassionate distribution of the fruits of economic growth (DESA, 2006), evoking good governance as key to protect and include the least well-off within a notion of universal solidarity.

The triad framework of Fainstein's *Just City* (2014) encompasses the range of concerns described earlier. Social justice is to be achieved through the accommodation of different and perhaps conflictive interests of *democracy*, *equity* and *diversity* aiming for as close as possible to equilibrium. *Democracy* focuses on the policymaking process, addressing not the *ends* but the *means*. It regards the fairness in circumstances and how participatory and transparent decision-making is; in other words, how citizens of different backgrounds and identities have a voice or feel their interests are adequately represented (Parnell & Robinson, 2006) in important deliberative arenas (MacLeod & McFarlane, 2014; Nikšič & Sezer, 2017).

Equity considers the different (normally economic) starting points of individuals, trying to bridge the gap by counterbalancing the push for growth and accumulation, through a fair distribution of wealth, resources, benefits and opportunities (Nikšič & Sezer, 2017). Equity prompts planning considerations, such as who has access to public space? And for what purposes can public space be used? (Fainstein 2009: 29–30).

Diversity, finally, involves the recognition of and respect for group differences (Young, 1990), embracing cities' multiplicity (Nikšič & Sezer, 2017). It acknowledges identity grounds for oppression and injustice, including class, race, gender, ethnicity, sexuality, ability and age, which intertwine with objective material distributional matters (MacLeod & McFarlane, 2014). Fairness through diversity means the *inclusion* in city spaces of the different 'identity' groups where they feel legitimate (Young, 1990).

Since their appearance, markets have been a democratic space with possibilities for attaining social justice. However, this potential is being jeopardised, as markets are currently under threat: cities, markets and local governments around the world flourished on a symbiotic relationship established around urban supply (Janssens & Sezer, 2013), but the advent of more modern provision facilities (as supermarkets), competing on an unequal footing with traditional markets, led to their decay. Municipalities gradually stopped supporting these infrastructures considered outdated and unnecessary (e.g. Mehta & Gohil, 2013), while other policy agendas (e.g. housing, education) were prioritised (Watson, 2006). Social (justice) impacts of markets' decline include, among others, diminished employment provision and social integration.

In an attempt to reverse markets' deterioration and the *laissez-faire* attitude of local governments, recent academic studies (e.g. Morales, 2009; Watson, 2009) and institutional initiatives (e.g. Urbact Markets—European Union) have advocated for their preservation. Markets are defended for having plural functions, which can be directly linked to Fainstein's triad (2014):

1) The planning for markets should involve the *democratic* opportunity for participation of all the groups of stakeholders impacted by decisions. 'Market communities' (here loosely defined as people interested in and routinely engaged with specific markets) can be prompted to claim the right to engage in political decision-making around the governance of these infrastructures.
2) In terms of *equity*, markets promote food security, and their low initial capital investments are important in enabling livelihoods (González, 2020). Trading at markets makes upward mobility possible while empowering traders (Morales, 2009). Moreover, markets play an essential role as public spaces where social and cultural practices happen (e.g. Pereira, 2017), and stakeholders establish relationships within the community (Freire, 2018; Urbact Markets, 2015).
3) Markets simultaneously provide a sense of local authenticity (Mehta & Gohil, 2013) and settings of a supralocal familiarity, where routines and codes of conduct are intrinsically known (ÜnlÜ-YÜcesoy, 2013). People from diverse backgrounds tend to feel equally welcomed at markets; hence, their contribution to cities' experience of *diversity* and integration (Schappo & van Melik, 2017).

Municipal markets can fulfil urban functions aligned with social justice aims, even if policy discourses do not explicitly state this. Still, planning can be a key catalyst as, although commonly inclusive, markets do not automatically embody progressive ideas. Local governments' promotion of social justice through markets' functioning is observed and defended by, for example, Morales (2009). However, while literature reveals local councils are (to some extent) aware of markets' multiple functions, academic analyses of contemporary planning for markets often convey a strong criticism of their drivers and/or effects. Cases across the world (e.g. Freire 2018, González, 2020; Pereira 2017) present similar governance scripts driven by a neoliberal rationale, where markets are commodified through privatisation or outsourcing. Revitalisation projects performed by private partners are defended and legitimised by discourses depicting markets as outdated, decayed and empty infrastructures, with the onus of improper governance remaining with less well-off traders and users (Pereira, 2017). Renewed markets commonly become a simulacrum of the original ones (Freire, 2018), with sanitised, securitised and exclusionary spaces targeting wealthier users, producing gentrified markets (González, 2020).

Despite these critiques, cities' governance realities resist simplified interpretations. Besides municipalities' dilemmas about what agendas to prioritise, pragmatic constraints such as limited time, staff and budgets need to be factored into planning for policies' enforcement. Public managers' decisions may be driven by the wish to deliver fairer societies (Parnell & Robinson, 2006), but their actions and ability to control outcomes are limited. In Belo Horizonte, as in other metropolises, managers are constantly pressured by tight budgets and the often-contradictory aims to simultaneously promote economic growth, update or build infrastructure and provide basic services to vulnerable populations (Parnell & Robinson, 2006). Within this complex landscape, employing private management and capital in the revitalisation of markets, while economically driven, may be the only viable alternative to bring life back to markets condemned to abandonment (Freire, 2018) and to maintain—or even promote new—social agendas. The case of Belo Horizonte markets' management change exemplifies the challenges of accommodating multiple (and sometimes conflictive) governance agendas.

Markets governance and social justice: Belo Horizonte's case

Belo Horizonte is Brazil's sixth-most populous metropolis, with 2,521,564 inhabitants (IBGE, 2020). Despite stark economic inequality, it is known for some innovative planning initiatives addressing the urban poor. The pioneering and nationally awarded food security *ABasteCer* (Supply) Programme is particularly relevant here. Created in the early 1990s, it addressed food deserts in peripherical low-income neighbourhoods of the city through the establishment of big warehouse-like shops called '*sacolões*' ('big bags'). These commercialise produce in bulk very cheaply, with associated shops required to sell 20 items at a low price determined by the municipality. The *ABasteCer* was a huge success, and units spread through the city. As unbeatable suppliers, *sacolões* regulate fresh produce prices in Belo Horizonte, and the decline of public markets' provision relevance in the city is intrinsically connected to their appearance.

While the 1990s saw a local government committed to social agendas, since the election of Marcio Lacerda (2009–2016), governance follows an entrepreneurial rationale, and in 2011, the municipally owned company PBH Ativos was created to give specialised technical support in public-private partnerships and tendering projects.

Belo Horizonte and its marketplaces

Belo Horizonte's population has an emotional bond with places of traditional trading, as interviews with locals revealed. Nevertheless, only four of the seven permanent enclosed marketplaces the city once had under its management are still functioning. The first market of Belo Horizonte, Mercado Central, is economically profitable and very lively: a democratic centralising place, reuniting all sorts of people, including tourists. It is, however, no longer public: it has belonged to a cooperative of more than 500 traders since 1964. This case contrasts with the public markets' reality of decline. Among the six remaining markets, two have been shut down. The first was demolished in 2011, giving way to a private hospital (Mercado da Barroca), and the second has been empty since closing in 2007 (Mercado de Santa Tereza). The Mercado da Lagoinha, moreover, was converted into a food security training centre.

The remaining three public functioning markets have very different scales and economic conditions. On the one hand, there is the large-scale Mercado do Cruzeiro (Figure 9.1): located in a wealthy neighbourhood; it houses 55 traders and is economically stable. On the other hand, there are the smaller markets, FECOPE (Figure 9.2) and CAM (Figure 9.3). The first is in a middle-class area, and the second is surrounded by low-income neighbourhoods, for whose residents it still has an important food supply function. Both of these markets nowadays have at least half of their units vacant. Figures 9.1–9.3, from 2019, portray these markets' pre-COVID-19 realities.

Figure 9.1 Mercado do Cruzeiro, Belo Horizonte.

Source: author.

Figure 9.2 Corridor of empty shops at FECOPE, Belo Horizonte.
Source: author.

Figure 9.3 CAM with some vicinity customers.
Source: author.

Markets current and future governance

Belo Horizonte's public markets are managed by the SUSAN within the SMASAC. This institutional position reveals that despite the very limited scope, markets are still considered food supply infrastructures. Nevertheless, they occupy a peripheral place within SUSAN's programmes. For over 20 years no meaningful measures were taken to improve management approaches to tackle the increasing decline of markets, partially caused by SUSAN's very own *ABasteCer* programme. The selection of market traders through a bidding system is the only significant governance element performed by SUSAN nowadays. This system determines a minimum rent price for each shop unit, over which traders make a bid. As long as they meet basic requirements (like clear debt records) with the municipality, the candidate offering the highest amount wins. Selected traders can occupy the units for up to five years, with contracts renewed yearly.

It is legally documented that merchants have *precarious rights* in their official enrolment with the city council, which is further emphasised by the municipality's right to request units back at any moment. In the process of interviewing traders, complaints about the system were commonly voiced, mostly about the constant fear of losing 'their' shops in each new bidding round. Established traders are given no preference in the application process, a condition defended by a council manager as essential to maintaining the impartiality and fairness of the selection process. Conversely, ad hoc practices show that SUSAN staff members commonly turn a blind eye to traders' informal practices (for instance, sub-renting shop units) to cope with constraining contract conditions, revealing their empathy and respect for those whose livelihoods depend on the markets.

After years of poor municipal engagement with marketplaces' governance, the new mayor, Alexandre Kalil (2017 onwards) brought this to a higher sphere of decision-making. An interdisciplinary workgroup (WG) reuniting eight heads of secretariats was created in September 2017 to discuss alternatives for the markets' management and renovation. The range of WG secretariats including, among others, treasury, planning, culture and social services indicated an understanding of markets' plural nature and functions, in line with academic discussions (e.g. Morales, 2009). Following the entrepreneurial governance tendency of recent years, the WG determined in 2018 that the markets should be tendered, and this outsourcing process was organised with the technical assistance of PBH Ativos up to 2021. The aim was to transfer the direct responsibility and costs of managing and refurbishing the markets to aspirant 'concession holders' (the private partners selected through the tendering). These chosen partners, in exchange, can economically exploit markets for 25 years. In terms of ownership, the markets remain public and institutionally under SUSAN's direction, which now *monitors* the concession holders' compliance with the mandatory tendering terms.

The rationale for adopting this strategy, rather than direct investment by local government, was based on two pragmatic arguments, repeated by different WG participants. They represent the council's understanding of markets' present situation and contribution to the city: (1) markets are emptied out of public functions: poor in supply relevance, cultural identity and even socialising significance, therefore 'only' serving traders' individual economic interests; (2) the municipality has more pressing social agendas to cover with public funds.

Despite the grim picture, the WG's motivation to find governance solutions for the markets is embedded within a more positive context. The feeling that markets are important elements of the laid back, friendly and gastronomically rich *Mineira* culture, whose preservation is paramount, is exemplified in the activism performed by neighbourhood associations for the Mercado do Cruzeiro and even the Mercado de Santa Tereza, despite it being shut since 2007. The bottom-up debates prompted by the associations reached the city scale and were important drivers for the creation of the WG, which agrees that markets are, beyond commercial spaces, the locus of employment, socialisation, culture and traditions. Moreover, SUSAN staff declared their belief that markets can be improved, through the tendering, and gain back supply functions; however, with a different emphasis on agroecological and organic goods from producers of Belo Horizonte's metropolitan region.

The chapter now examines the extent to which the government's representation of markets is (consciously) aligned with social justice promotion in the city and its further impact in their planning. Rather than simply addressing whether or not the municipality (intends to) support(s) the markets and why, it analyses *how* the WG members decided to operationalise this support. The tendering phases and contractual conditions reveal the governments' vision for the newly managed markets and provide answers to concerns related to which stakeholders benefit or are burdened by this governance change, which social groups will access and feel belonging in the new markets, what functions markets will fulfil and what priorities drive tendering decisions. For analytical purposes, the elements of the tendering are now divided between Fainstein's (2014) *democracy*, *equity* and *diversity* categories while recognising that there are constant overlaps and, conversely, trade-offs between these axes.

Markets' tendering and social justice

Democracy

The tendering development process included three opportunities for public participation in the planning stages. The first, mandatory, was a *Procedure of Interest Manifestation* (PMI). The PMI invited interested stakeholders to submit proposals for the markets. The expected contributions could be made to any, a few or all of the public marketspaces described previously. Obligatory guidelines had to be respected as the *"preservation of markets' typical activities"* and the need to *"consider the socioeconomic and urbanistic conditions of the markets and surrounding neighbourhoods"* (PBH Ativos, 2018). Participation in the PMI did not guarantee any preference in the future tendering selection, and submitted proposals became the property of the city council.

The second participation opportunity happened because a required law allowing the tendering was seeking approval in the aldermen's chamber (CMBH).

It was a public session promoted by the president of the *Human Rights Commission*, a left-wing alderman who wanted to clarify tendering terms and intentions. Besides WG members, traders' representatives were invited to this session, while anyone could attend. The last participatory element within the tendering was the consultation phases about the two released tendering calls. Each consultation was open for a month, inviting feedback from any interested citizen. The municipality provided answers to all submitted feedback, and some input was incorporated into the final calls.

In terms of democracy, despite these opportunities for public participation in the tendering, it required availability, a certain level of technical expertise and even economic resources of the public. Consequently, the active participation of less-educated and less-mobilised stakeholders—such as the traders of FECOPE and CAM—was not feasible. Conversely, in cities' uneven fields of power, the most important is not participation per se but to have interests fairly represented (Fainstein, 2014). The results of public policies are therefore more important than the circumstances under which they were created. The WG members expressed responsibility towards traders' main concern: secure tenure within the markets. Hence, concession holders are legally obliged to allow merchants to continue trading in the markets while paying the same current rent fees for five years.

Equity

Belo Horizonte has pronounced economic inequalities, and therefore a strong justice needs to address equity. Further imposed tendering conditions in line with equity concerns and economic inclusion are as follows:

1) The joint tendering, coupling potentially profitable markets with less economically interesting ones. FECOPE and CAM were not considered viable investments by PMI proponents, and their future was enquired about in the public session. Influenced or not by these event discussions, the WG grouped together FECOPE and the Mercado de Santa Tereza, and CAM and the Mercado do Cruzeiro. The strategy clearly imposes a trade-off to concession holders.
2) To provide 25% of each market's units for half of the normal rent fees.
3) To destine at least 30m², also at subsidised prices, for an 'agroecology store' aimed at promoting healthy food intake, sold at competitive prices by small producers linked to urban and family agriculture programmes (PBH, 2020).

The literature (e.g. Morales, 2009) defends markets' relevance for employment due to low entry barriers. While the previously listed measures do safeguard opportunities to 'lower income' traders, a way of promoting equity in cities (Fainstein, 2014), the tendering of markets is not a social service project. The economic viability of the markets is a basic premise of the tendering. WG members stated that "*the selling of oranges won't sustain the markets*", hence the importance of businesses like restaurants working as 'anchor enterprises'. Still, the listed requirements can serve as partial shields to gentrification. Although not designed specifically to contain it, the conditions provide evidence that the council members *"don't want shopping malls"* as the tendering result. The incorporation of social agendas also indicates the WG's tendering employment as a rather neutral management strategy, instead of an instrument to commodify public assets. According to SMASAC's vision, it is the most viable way of enabling public policies in relation to the markets. As such, private capital would be at the service of the public authority's agenda rather than the social conditions serving as a rhetoric discourse covering profitability priorities.

Notwithstanding, when assessing what is the priority in the tendering process, one has to remember that the city council gains three times in the new arrangement: it is no longer responsible for employing public funds to support markets; it can allocate fewer personnel to monitor markets' management and, importantly, it collects revenue with the tendering. The selection of concession holders happens in a similar fashion to the bidding of shops, although at a much grander scale: annual basic fees are established and the candidates offering the highest bid win. In response to enquiries in the consultation phase about the adopted price criterium, the municipality argued that the choice does not mean privileging earning over preserving markets' typical nature and functions. Since the tendering contracts are clear about mandatory terms, revenue collection is a secondary reasonable advantage which the city council will retain. Further supporting the price criterium, a SUSAN staff member explained that it is the most pragmatic one for the public authority since it is less subject to contestation about favouring specific candidates.

Diversity

Regarding diversity, the vision for the new markets expects lively, diverse and culturally rich places, with an increase in numbers and plurality of traders and customers. The investment options and branches cover opportunities targeting from small agricultural producers needing subsidised spaces to highly capitalised entrepreneurs. Moreover, the expected wider catchment area of markets, together with the different types of activities hosted, aims to attract users with varying social and economic profiles. The establishment of gastronomic and cultural activities also aims to attract tourists, a type of public not currently seen in the public markets.

An important requirement within the tendering conditions supporting these aspirations is the creation of a social committee. Formed by representatives of the city council, who will preside over it; the concession holder; the *Municipal Council of Food Security and Nutrition (COMUSAN)*; the traders; and the neighbourhoods' associations, the group has to meet at least bimonthly and "elaborate exclusively social guidelines, for the fulfilling of the intrinsic matters related to the tendering" (PBH, 2020—Annex II: 37). While the social committee will have limited functions and powers, the forming of such networks of stakeholders will potentially be instrumental in the construction of social capital resources associated with the markets. This requirement indicates the municipality's intention to promote a broader, more diverse and sustained civic participation in the process of monitoring the markets' future functioning, such as the social impacts that the new management system will generate for market traders and users.

Among the concerns, the committee can help address the risk of expulsion of specific vulnerable groups from new markets' environments. Concession holders will be catering for more capitalised users with sanitised and closely controlled markets, where customers' expenditure is expected to increase. Encroachment on this space by informal traders and homeless dwellers, seen in the context of FECOPE and CAM, is very unlikely to be permitted. Thus, markets' functions as public spaces are to be partially compromised.

Conclusion

The governance of Belo Horizonte's public markets presents a common contemporary situation faced by marketplaces around the world, with the (partial) transfer of management responsibilities to the private sector (Pereira, 2017). Still, beyond the reasons for choosing a tendering process, the application of this strategy to Belo Horizonte's markets provides some public governance and research insights, in light of social justice intentions: the obligatory terms within contracts and the vision for the refurbished markets.

While the depiction of markets' current functions as limited to traders' livelihoods is a reductionist portrayal, joined with governments' apparent budget limitations, such representation is instrumental to legitimise the tendering choice instead of public investment. Moreover, the advocacy performed by the heads of SUSAN and SMASAC, with the credibility invoked by their institutional position, aligned with social agendas, strengthened the idea that the tendering is a reasonable and neutral, rather than a political governance strategy. Although acknowledging the entrepreneurial logic, the WG adopts the tendering as the viable solution to stream private resources to support markets, while still directing the outcomes.

Some of the tendering contractual terms indicate the council's awareness of markets' multiple functions for the city. Inspired by the local fondness for traditional trading forms, and the previous activism around markets' preservation, WG members declared their personal interest in seeing markets living up to their apparent potential. Therefore, the implementation of the tendering process, since the creation of the PMI, reveals the municipality's effort to reasonably equate markets' indispensable economic sustainability conditions and the desired social results. Social (justice) interests are perceived in the secure tenure of current traders, the creation of a plural social committee, the trade-off of coupling a profitable market and one more likely to be subsidised and the percentage of units to be rented at lower prices, with the obligation to sell sustainable agroecological goods.

Yet, there are limits to the extent to which the new management and the vision for the new markets coincide with social justice aspirations. As the analysis employing Fainstein's triad (2014) revealed, participation was limited, despite the opportunities offered, and the sort of diversity to be expected in the rein-vented markets' environments appears to concentrate around middle- or upper-class user groups. Still, each of the four tendered markets will develop its own dynamics, and gentrification will occur (or not) in undoubtedly different ways, depending on markets' size and specific neighbourhood.

Marketplaces can be employed as part of a planning strategy to secure higher levels of urban equity and justice, but they do not automatically embody progres-sive ideas. Thus, here it is essential to draw attention to the crucial role played by council managers in charge of monitoring the functioning of markets and their new managers. While mandatory conditions in line with social justice are estab-lished in contracts and law, these conditions have to be effectively enforced. Capitalist logic otherwise may strongly jeopardise any of the 'progressive' ten-dering terms detailed in this chapter. At the moment, with the material results of the tendering still to be seen, one might assume that managers' intentionality is driven by good governance principles.

Belo Horizonte's example reveals the complexity of cities and their govern-ance realities, where adopted planning solutions try to balance growth and equity (Fainstein, 2009) interests in creative ways. The analysis of tendering as the solution to support markets, therefore, reveals it not to be a *"blind mimicking of a neoliberal agenda"* (Parnell & Robinson, 2006: 351). Even if employing a typ-ical neoliberal strategy, the city council reveals awareness about markets' (poten-tial) plural contribution to the city, representing markets as infrastructures where entrepreneurial and social justice aims can coexist.

References

Department of Economic and Social Affairs – DESA/UN, (2006), *Social Justice in an Open World: The Role of The United Nations*. New York. United Nations.

Fainstein, S. (2009), Planning and the just city. In: Marcuse, P., Connolly, I., Novy, J. et al., (Eds.), *Searching for the Just City: Debates in Urban Theory and Practice* (pp. 19–39). New York: Routledge.

Fainstein, S. (2014), The just city. *International Journal of Urban Sciences*, 18(1), 1–18.

Freire, A.L.O. (2018), Mercados Públicos: de equipamentos de abastecimento de alimen-tos a espaços gastronômicos para o turismo. *Geografares*, January–June, 176–198.

González, S. (2020), Contested marketplaces: Retail spaces at the global urban margins. *Progress in Human Geography*, 44(5), 877–897.

Harvey, D. (1973), *Social Justice and the City*. Athens: University of Georgia Press.

Hayek, F. (1979), *Social Justice, Socialism and Democracy: Three Australian Lectures*. Turramurra: CIS.

IBGE (2020), Belo Horizonte [online]. *IBGE*. [Viewed 7 September 2020]. Available from https://cidades.ibge.gov.br/brasil/mg/belo-horizonte/panorama

Janssens, F. & Sezer, C. (2013), 'Flying markets' activating public spaces in Amsterdam. *Built Environment*, 39(2), 245–260.

MacLeod, G. & McFarlane, C. (2014), Introduction: Grammars of urban injustice. *Antipode*, 46(4), 857–873.

Mehta, R. & Gohil, C. (2013), Design for natural markets: Accommodating the informal. *Built Environment*, 39(2), 277–296.

Morales, A. (2009), Public markets as community development tools. *Journal of Planning Education and Research*, 28(4), 426–440.

Nikšič, M. & Sezer, C. (2017), Public space and urban justice. *Built Environment*, 43(2), 165–172.

Parnell, S. & Robinson, J. (2006), Development and urban policy: Johannesburg's city development strategy. *Urban Studies*, 43(2), 337–355.

PBH (2020), Concorrência N° 001/2020 [online]. *Prefeitura Belo Horizonte*. [Viewed 7 September 2020]. Available from https://prefeitura.pbh.gov.br/fazenda/licitacao/concorrencia-001-2020-1

PBH Ativos (2018), PMI Mercados Municipais [online]. *PBH Ativos*. [Viewed 7 September 2020]. Available from http://pbhativos.com.br/pmi-mercados-municipais/

Pereira, C.S. (2017), Mercados Públicos Municipales: Espacios De Resistencia Al Neoliberalismo Urbano. *Ciudades*, 114, 39–46.

Schappo, P. & Van Melik, R. (2017), Meeting on the marketplace: On the integrative potential of The Hague Market. *Journal of Urbanism: International Research on Placemaking and Urban Sustainability*, 10(3), 318–332.

ÜnlÜ-YÜcesoy, E. (2013), Constructing the marketplace: A socio-spatial analysis of past marketplaces of Istanbul. *Built Environment*, 39(2), 190–202.

Urbact Markets (2015), *Urban Markets: Heart, Soul and Motor of Cities*. Barcelona: Institut Municipal de Mercats de Barcelona (IMMB).

Watson, S. (2006), *City Publics: The (Dis)Enchantments of Urban Encounters*. New York: Routledge.

Watson, S., (2009), The magic of the marketplace: Sociality in a neglected public space. *Urban Studies*, 46(8), 1577–1591.

Young, I.M. (1990), *Justice and the Politics of Difference*. Princeton, NJ: Princeton University Press.

10 Lima markets beyond commerce

Challenges and possibilities of common food spaces in periods of crisis

Ana María Huaita Alfaro

Introduction

This chapter aims to explore marketplaces in terms of their potentialities as spaces of common urban purpose rather than prioritising individual and private interests from commercial exchanges. These potentialities are analysed from observations on encounters and experiences around everyday practices of food provisioning at marketplaces. The study focuses on aspects that surpass commercial activities and takes a closer look at interpersonal exchanges. Encounters, as forms of co-presence in a shared space, allow the unveiling of common urban purposes that become expressed in the capacities of individuals to negotiate divergences and differences for being together and living in common.

The study looks at the city of Lima, a context of great particularity, on the one hand, because of the continuity and growth reported for the retail market sector. On the other hand, the city and residents' forms of common living have been frequently evaluated in terms of weakening civic ties that become expressed in spaces and manifestations of the public realm. This context raises the interest to observe how markets, as spaces of daily encounter and interaction, may prevail on top of urban conditions appearing to challenge residents' capacities to negotiate togetherness and conviviality.

In contrast to other capital cities in Latin America, where supermarkets have taken the lead in food provisioning and occupation of the urban food space, Lima markets still have an important role in providing food supplies city-wide. According to official estimates, Lima's population represents a third of the Peruvian population—nearing ten million inhabitants (INEI, 2014). To serve this large number of residents, Lima concentrates 43% of national markets, equal to 1,122 markets, in varied shapes and sizes—old and emblematic markets, neighbourhood markets, street markets, managed by municipalities or traders' associations. There are over 2,600 markets in the whole country (INEI, 2016).

In 20 years, from 1996 to 2016, markets have doubled in number at the national level. Supermarkets have also multiplied around six times in this period, but these do not supply the larger national and urban majorities. The great majority of residents still rely on markets and other forms of direct commerce (street vendors and small shops) for their provisioning (INEI, 2016). Markets remain the primary source for over 60% of the national population (FAO, 2016).

DOI: 10.4324/9781003197058-10

Despite their relevance, marketplaces are often found as disregarded infrastructures for basic supplies—sociocultural centralities often left behind in the current configuration of the modern city (Tello & Narrea, 2014). Moreover, city residents use and engage with today's markets as visitors and unattached consumers, as if conceiving the urban space—and the city as a whole—as strangers' territory, reflecting a lack of identification with what constitutes public and common dimensions in urban dwelling (Huaita-Alfaro, 2019).

In line with these observations for Lima, studies on residents' use of the public realm, such as markets, point to increasing individualisation and disengagement with them (Joseph et al., 2008; Protzel de Amat, 2011). As Díaz-Albertini (2016) argued in his study on Lima public spaces, the reliance on free-market mechanisms to address urban demands has undermined avenues for participation and collaboration among the urban collective. Urban development projects evolve against the availability, accessibility and qualities of spaces for encounter, leaving behind opportunities to collectively work towards residents' needs and aspirations from the city.

Urban divisions are stressed in periods of social crisis, such as that currently experienced due to the pandemic of COVID-19, impacting on global health systems and bringing about unforeseen multidimensional impacts to urban communities worldwide. As physical proximity is particularly avoided given the nature of this health crisis, discussions on togetherness, living together and conviviality tend to be conflictive. Nonetheless, spaces for providing essential services have not stopped operating and offering daily chances of commingling, uncovering a pending agenda for rethinking and re-engaging with modes of living together.

In this chapter, I build my analysis upon market explorations under two periods. The first period corresponds to my years of PhD research on inner-city markets of Lima between 2013 and 2018. During this time, I conducted a six-month fieldwork period, enriched by an extensive revision of secondary sources of information on the quotidian reality of national and international markets. I worked in two inner-city markets during this fieldwork, where I applied ethnographic methods and conducted 36 interviews with a group of traders, customers and municipal authorities, in addition to other meetings conducted with researchers and specialists. My research focused on analysing market encounters among main market users, traders and customers, taking food as a lens to explore how encounters took place and what these may unveil about the construction of markets as common spaces (Huaita-Alfaro, 2019).

The second period corresponds to the first year of the COVID-19 crisis in 2020. My engagement with multisectoral initiatives addressing urban food markets allowed me to expand my previous work. These initiatives included a capacity-building project directed to market traders and leaders of the National Federation of Market Workers in Peru, with the support of national and international organisations (mainly Friedrich Ebert Stiftung Foundation in Peru, ECOSAD NGO and allies such as the Ministry of Production of Peru, Municipality of Lima, FAO Peru). Likewise, I engaged with multisectoral platforms on food systems and environments promoted by FAO Peru and the Municipality of Lima—for example, the Roundtable for the Promotion of Healthy Food

Environments Policies. My participation in these spaces allowed me to approach the problem of markets and how they secured their continued operations during the health crisis, based on the narratives from traders and customers throughout the discussions promoted by these initiatives. Furthermore, there has been the chance to present and enrich information from this period by sharing the outcomes of the research in national events and publications (Huaita-Alfaro, 2020; Santandreu et al., 2021).

The following sections will present an approach to urban markets beyond the regular accounts of their commercial roles. Challenges from individualisation and privatisation logics at markets will be introduced to then move on with presenting possibilities that emerge from daily actions and negotiations taking place at markets. Overall, both research periods show that markets actors are continuously acting towards the possibilities of sharing markets as common ground. Moreover, the pressing need for collective responses during periods of crisis can strengthen the sense of common urban purpose continuously evolving in these spaces.

Approach to urban markets beyond commerce

Marketplaces have been historically at the centre of cities' political, economic and social organisation. These spaces have played key roles in shaping urban configurations throughout the consolidation and growth of cities. In turn, the configuration and distribution of markets portray interesting aspects of social and spatial transformation. In that sense, marketplaces are living spaces that reflect how a society experiences urban processes. This reflection occurs through collective and individual actions of markets' appropriation and use, and through the multiple encounters taking place alongside, as De La Pradelle (1995) argued.

By reflecting the 'rhythm' of society (see also Chapters 2 and 3), these spaces result in constant adaptation and transformation, responding to the dynamics of urban life (Medina & Alvarez, 2009; Medina, 2013). These characteristics have made markets prevail despite pressures on economic growth from ruling political and economic systems, urbanisation forces and changes in labour and consumption conditions favouring private, virtual and globalised forms of exchange (Pottie-Sherman, 2011). Moreover, in their traditional and emblematic forms, markets can be recognised for their qualities as permanent spaces of food provisioning and open to the diversity of urban publics, keeping these as spaces for everyday commingling and everyday contestation of collective challenges in urban living (Buie, 1996).

The openness of market settings and exchanges provides occasions in which it is possible to participate in moments of co-presence around shared actions and interests over the production of common space (Simone, 2014). Thus, market operations can be evaluated as shaping the physical settings and implying forms of active or absent participation in the functioning and continuation of these sites (De Certeau et al., 1998). These operations unveil how people deal with sources of urban conflicts, such as privatising logics leading to the loss of shared, collaborative and convivial forms of inhabiting the urban space (Amin, 2008).

In such a way, operations of production, consumption and exchange at markets are regarded as presenting avenues for overcoming sources of urban fragmentation.

Furthermore, sociocultural significances applied to marketplaces—for instance, concerning their functions and populations—define residents' relations with the urban space and how they place themselves in relation to those with whom they share this ground (Fincher, 2017). These significances are not fixed but as fluid as residents' experiences and trajectories in the city that work to define residents' differentiated standings in urban relations and their recognition as active agents in city making and living (Fraser, 2000). Under these notions and views on urban spaces, individuals and other agents impact the everyday making of the city.

Following these considerations, encounters with the urban space and its diversity of users are enlightening experiences from which to raise outcomes from the shared actions and interests meeting in the everyday production of markets. Encounters, as individuals' intersections in a shared space, take place through social and material arrangements and allow for expressions of convergence or divergence among the diversity of urban actors (Fincher & Iveson, 2008). Moreover, materialities taking part in these encounters are useful in exploring tensions or contestations that may revolve around aspects of common purpose (Koch & Latham, 2012). This is the case of food, around which people relate while exchanging modes of experiencing the city (Watson, 2009). Thus, relating around converging food interests may bring about expressions of negotiated recognition and participation in the urban space. Hence, food is a useful and enlightening means to deepen the approximation of markets' local environments (Huaita-Alfaro, 2019).

By addressing marketplace encounters and daily experiences, the intention here is to address how the common space allows for actions over shared purposes and resources beyond economic aims. Through actions such as collaboration and co-creation, market users can negotiate urban conditions and add efforts for enhancing markets' management and functioning (Higuchi, 2015). Avenues for such converging actions will be explored in the following accounts on Lima. Moreover, these lead to identifying the need for effective arrangements—addressing political, economic and social dimensions of markets—for such convergence to happen, connecting the diversity of market users and urban actors in the continuation of these common spaces (Simone, 2014).

Markets for commingling: Challenges from commercial operations

Markets in Lima are easily encountered when transiting along the city, gathering large numbers of diverse urban agents in and around markets (Filgueiras, 2009). Despite their important presence, markets are often found as neglected spaces, both for physical and social considerations (Tello & Narrea, 2014). The situation for markets can be evaluated within contemporary processes of urban growth, led by political economy arrangements moved by a privatising imperative shaping urban governance and city making (Babb, 2008). Likewise, it is important to acknowledge local processes of social order, strongly marked by social differences and weakened civic relations that threaten the continuation of spaces of common use and appropriation (Joseph et al., 2008; Protzel de Amat, 2011).

The first period of study importantly informed on the long-lasting problem of Lima markets. In market users' accounts, there is a clear reference to contextual issues from what they could closely observe in their commercial and consumption experiences. Users mainly pointed at concerns on infrastructural maintenance and the modernisation of commercial operations that have not favoured markets' prevalence against private and modern spaces of commerce. Ines, trader of Mercado N.1 in Surquillo district, mentioned in an interview how markets dealt with the appearance of supermarkets, and how conflicts in their management, reflected in infrastructural conditions, seem to be hard to overcome:

> [W]hen the economy changed [under neoliberal measures] in Peru, more [super]markets started to appear.... Metro, Wong, Vivanda appeared. All of us here [traders] had a bit of fear, but little by little we have lost that fear and we have made it work because we have a fixed clientele. But people [traders] should thank God because customers come in. (…)
>
> (Ines, trader, Mercado N.1 in Huaita-Alfaro, 2019)

This context for markets can be explained in relation to their privatisation, taking place along a political economy transition from public to private imperatives and modes of urban governance (Babb, 2008). In the early 2000s, a privatisation law for municipal markets was published that encouraged traders to associate and buy market lands and infrastructures, and other traders' collectives to develop new market projects (Peru. Law N° 26569, 1996). In this way, the large number of markets today found in Lima is a free-market response to labour demands on one side, and to demands on basic and accessible goods by a growing population on the other.

This transition to private market ownership and management took place along the institution of reduced municipal functions in the surveillance of privatised services, regardless of their public reach. Thus, public authorities increasingly left markets in hands of private and profitable interventions (Ludeña, 2010; Peters & Skop, 2007; Vega Centeno, 2014). Limited responsibilities from public actors have negatively impacted markets' health and safety conditions for city residents and undermined the open nature and collective services of these spaces.

Municipalities have maintained specific roles in the surveillance of market services, such as those related to food safety, but their interventions have actually been largely restricted to the extent to which markets prove to be profitable investments for the municipality. These views were raised by Octavia, head of the Economic Development office at the Municipality of Surquillo, in one of the interviews:

> She explained that because of bad relations with traders, and the little or no income they represent to the municipality [from taxes], the public administration undertakes the least possible actions to invest the least possible

money there. She said they were in charge of cleaning and providing security, and therefore there were some municipal guards in the area. But the minimum effort was put on these activities. If there were other types of failures, such as problems with electricity, they leave traders to solve these.

(Notes on Interview 1—Octavia, Municipality,
Mercado N.1 in Huaita-Alfaro, 2019)

Following privatisation trends, market users such as traders, customers and municipal authorities identify traders as the main or only actors in charge of undertaking improvements and in securing the continuation of marketplaces, regardless of the level of responsibility and difficulties implied in securing their public services of food distribution and accessibility, as well as the surveillance of market areas.

The privatisation of marketplaces and individualisation of management arrangements was not only promoted by public institutions but also supported by traders themselves, who trusted private solutions to be the most effective responses facing the absent interventions from public authorities in markets' protection and surveillance. The political and economic rules introduced for markets defined their forms of management, as well as the commercial context in which these operate, with traders participating as individual entrepreneurs (Muñoz & Rodríguez, 1999).

Even if conforming associations and becoming managers and/or legal owners of markets, a strong sense of individual competitiveness among these small entrepreneurs has undermined their collective efforts. Julia, another trader at Mercado N.1, mentioned,

Sadly, colleagues are not aware of the value of the market, of their stalls, they're only concerned about their own [sales] (…) each one looks after their own work, and if they have to give something to improve something [at the market], you have to ask them once, twice. They're so miserly.

(Julia, trader, Mercado N.1 in Huaita-Alfaro, 2019)

Private ownership and for-profit logics on traders and municipalities' operations at markets have been at the core of the scarce interventions maintaining these centres. Municipal and other public authorities have basically ignored plans for marketplaces' recovery. Municipal actors regarded markets as centres of political and economic influence, but not as suitable investments, as was made explicit by Octavia, representing the municipal authority in one of the case studies.

Moreover, urban growth dynamics have contributed to social fragmentation through their segregating impacts—for instance, in the distribution of urban services that reinforce social differences that also define relations among residents and their relationship with the varied city they inhabit (Joseph et al., 2008). In this regard, markets users interviewed during fieldwork generally evaluated these markets and their traders as not moving ahead in time. These views have also been intrinsically connected to notions of poverty and backwardness.

Karen, a customer at San Jose Market in Jesus Maria district, evaluated working at markets as linked to educational levels and possibilities to access other labours:

> I guess they have studied, primary and secondary school. I don't know if they have technical or university degree, I don't think so. I have the impression that most of them don't, otherwise they would be working in another thing.
> (Karen, customer, San Jose Market in Huaita-Alfaro, 2019)

Notions on markets and traders connect to everlasting tensions deriving from histories of migration and cultural mix in the city, likewise reflected in the diversity of users in these urban spaces (see Figure 10.1). This sociocultural mix of the city has given sense to the creole culture characterising Lima's food and culinary manifestations, which have been recognised in markets as centres portraying such valued diversity (Municipalidad de Miraflores, 2007). Despite the inclusiveness portrayed in the food imaginary, traders and customers approach the shared grounds of markets under different positionalities, different urban demands and possibilities to address them. Nidia, another customer at San Jose Market, reflected it in her views on traders' capacities and assumed short term visions on their work:

Figure 10.1 Mercado San Jose, Jesus Maria—Lima displaying the diversity of both products and visitors meeting daily, April 2015.

Source: author.

Maybe they're selling well now, they offer their products and they live with that, because they don't understand that improving certain aspects, they would sell twice they sell now.

(Nidia, customer, San Jose Market in Huaita-Alfaro, 2019)

These issues also point to absent interests in spaces for the public and the still missing involvement of private and civil society sectors in building synergies and negotiating common visions of the city. Urban governance frameworks have not been inclusive of residents' participation and have encouraged unfair competitiveness or favouritism in securing independent economic opportunities (Caravedo, 2013; Vega Centeno, 2014). Nonetheless, markets still provide commonalities for everyday commingling. Hence, addressing the prevalence of activities highly valued by urban residents, such as those around food, allows for evaluating ways of challenging governance operations, cultural constructions and civic actions impacting against interests, such as those around cities and food, connecting residents.

Crisis and response: Market possibilities as common urban grounds

The second study period was an important source of information on questioning how common interests still emerge on top of markets' challenging conditions for commingling. The public health crisis of COVID-19 has brought about significant struggles for traditional spaces of everyday encounters, precisely because it called for restrictions on the forms of togetherness and interpersonal exchanges at the core of its functions.

The context raised attention to spaces such as markets given the large attendance that these showed despite restrictions and called for interventions to address their continuity because of their vital roles in serving residents' needs. The introduction of sanitary and social distancing measures likewise showed markets' long-lasting problems in management and operation, which complicated the implementation of preventive measures and raised concerns about markets as potential centres of contagion.

This was also the case of Lima marketplaces, which continued operating mainly for activities of food supply (see Figure 10.2). Through images and testimonies shared on media and by national authorities, markets became the centre of public discussions addressing infrastructural deficiencies, while also raising missing commitments in caring for the physical spaces and essential services provided there (Páucar, 2020). This discussion has not been an isolated situation for Lima and other cities of Peru and can also be observed in other Latin American cities with an important presence of retail markets, such as Quito, Bogota and Mexico (Collyns et al., 2020).

In Peru, regulations published in response to the health crisis have given room to discuss their adequacy among decision-makers and urban actors related to markets' functioning. The first alarm on markets as sites of contagion made many of them temporarily close and reopen with strong adjustments. The following publication of emergency regulations made many market stands close for long periods because these were not evaluated under the scope of 'essential services'

Figure 10.2 Mercado El Eden, San Luis—Lima continued operating during the
       COVID-19 crisis by adapting facilities and services, April 2020.

Source: author.

or because regulations were not formulated by considering the large diversity of
conditions in the significant number of markets in Lima and the country. Nonetheless, the size of the market workers population and the size of the population
served by these centres made it unavoidable to question the substantial limitations on their activities.

   This concern on the side of traders was raised by the leader of the National
Federation of Market Workers in Peru in one of the activities organised as part
of a capacity development project with social organisations:

> COVID-19 has treated us badly. (…) There are questions from our fellow
> traders; they don't understand. In this situation, there have been many
> norms, many policy instruments that often confuse. If we don't understand
> them, imagine them [population of traders] (…) [by early July 2020] many
> fellow traders have not worked for over 100 days. We are worried because
> they want to return to work but don't know how; norms are not clear.
>   (Ricardo Ramos in Friedrich Ebert Peru Foundation, 2020)

The context raised traders' voices in their claim for market services but also
encouraged the involvement of urban, market and food systems specialists to

argue in favour of revising the markets' restrictions. These became public discussions, covered in diverse media, and increasingly gained influence in the re-evaluation of how markets were recognised and what roles made these sites endure, on top of the downsides claimed. New windows of dialogue opened across political and urban specialists, next to traders as field specialists. The Roundtable for the Promotion of Healthy Food Environments Policies, a participatory platform hosted by the Municipality of Lima, was one of the collectives raising their voices and intentions to collaborate in this regard:

> [Marketplaces] are a priority in the agenda of the different organisations conforming the Roundtable, we express our desire to join the actions to develop a new market system, seen as part of the agrifood system, next to municipalities and regional governments that should work to improve health conditions for consumers, traders and farmers, as it has been expressed by [Lima's] mayor Muñoz, concerning National Government interventions.
>
> (Consorcio Agroecológico Peruano, 2020)

Current regulations on markets functioning, considering sanitary and operational measures, are situated in this evolving understanding, focusing on preserving markets' roles in food provision and the subsistence of local economies. The norms have been updated after every evaluation of the emergency period (on average, after every three to four months). They particularly include sections and guidelines on market surveillance and promotion of their activities by municipal authorities. For this, municipalities receive an economic incentive to accomplish specific roles in markets of their jurisdiction (Ministry of Economy and Finance of Peru, 2021).

The Ministry of Production of Peru has been one of the main promoters in renewing commitments with the market sector. A series of projects in collaboration with this national entity has started over this period. It should be mentioned that these commitments have been mostly stressed after the installation of a transitory government between the end of 2020 and mid-2021, after a series of political changes in the country. Despite the short timeframe, it has brought about new lights on the revitalisation of markets. In this intention, the national government installed an Executive Roundtable on Marketplaces in January 2021, with the active participation of national leaders and directed by the minister of production, as he explained,

> There are two aspects to prioritise: to give an adequate space to traders to do their work in the most efficient way, and for citizens to get fresh, healthy and accessible food products in safe places. I will call a group of traders, experts in food chains, architects and designers to discuss what plan we can leave for food markets to be reformulated, redesigned in the coming years.
>
> (Jose Luis Chicoma in Villar, 2020)

Thus, market policies and strategies for their surveillance and promotion during the crisis have started to evolve as new priorities in urban agendas, and they have invited diverse urban agents to engage with their formulation, evaluation and practice. In Lima, the metropolitan municipalities next to traders' collectives and international agencies such as FAO are among the main actors joining in these renewed efforts. Likewise, there is an increasing involvement of public authorities with direct competencies in markets. The situation has also raised difficult dialogues among public sectors, as well as the need to deal with limitations in their capacities for action, other pending tasks in government proficiency to which authorities now have to engage with.

On top of the conflicts and concerns raised from markets' management and adaptation to the context, shared purposes among market leaders, decision-makers and other actors, such as opinion leaders on urban and food issues, have emerged as growing collaborations in the governance of local spaces. This encourages more concrete ways in which residents can participate in the good functioning of spaces of common use. For instance, urban leaders have joined under this purpose, encouraging dialogues and projects under the scope of markets' civic roles. This was expressed by the representative of UN Habitat in Peru as part of a series of events on social and labour inclusion next to urban collectives in Lima:

> [F]rom UN Habitat and working with Lima Como Vamos, we want to promote dialogues to see how markets can facilitate socioeconomic integration of the local population. We should understand that markets are not centres of contagion, but are centres that can facilitate exchanges under healthy distancing, maintaining economic dynamism and of course, generating productive chains and solidarity chains for the social equation.
>
> (Efren Osorio in Lima Como Vamos, 2020)

Evidence of market challenges raised by the current context demands further attention and reflection in the recovering of vital spaces for cities, bringing about opportunities for common living from key aspects such as political and economic participation, as well as social inclusion, around basic services for food provision. This remains an open discussion for essential community spaces, assimilating principles of equity, sustainability and food security that allow for healthy conviviality among residents and across urban centres and the complex ecosystems hosting them.

Conclusion

Operational concerns raised around markets in months of global upheaval have called for attention, not only on their sanitary conditions but also on the vital roles these urban spaces play for residents' everyday subsistence. Markets have shown a multiplicity of urban agents connecting around them and the need for nurturing engagement on their making and caring. From the information collected in Lima, market agents such as traders, consumers and municipal authorities have been observed as encountering stories that exceed the commercial performances of markets and the privatisation struggles that have shaped them. Instead, their current stories point to negotiations that allow them to secure their common interest, the sharing of the market space and the continuity of their activities.

There have been implicit collaborations among urban actors, particularly among traders, to keep these spaces alive. Still, concrete actions from other actors were missing, such as public authorities and customers inherently involved in their functioning. The crisis emerging from the COVID-19 pandemic has brought about opportunities, even from compulsory interventions, for explicit collaborations to start taking place among the diversity of agents linked to markets' continuity. Moreover, it has also fostered the need to re-encounter and recognise markets and users in their diversity of roles and contributions to provide viable solutions to still conflictive adaptations on current modes of operation.

Markets are centres of convergence for urban actors to collaborate in rethinking and remaking the city. Observations and reflections from this study point to avenues for the recognition of markets' roles in providing possibilities for residents' participation and fair access to the city. Market users' accounts informed about ways in which, beyond commercial activities, these sites allow for shared interests and benefits and from which a collective of agents can find room to collaborate and counteract the effects of private interventions in management and use. Explorations in this study raise the need of addressing how urban agents from different sectors bring about different views and forms of participation in markets' functioning while also building up relations that sustain the continuation of markets' services.

On the whole, it is worth acknowledging that the prevalence or recovery of common spaces is achieved because of collective actions impacting on their functioning and governance—from public authorities on their influential roles to residents in their everyday modes of use over these as shared resources. Participation in the making of spaces for living in common allows for the collective making of city politics that shape the built environment inhabited and experienced in common. From the challenges and possibilities raised, these sorts of local places may be envisioned as consistently contributing to largely pending gaps in local and national governance agendas for inclusive and resilient urban communities.

References

Amin, A. (2008), Collective culture and urban public space. *City*, 12(1), 5–24.

Babb, F. (2008), *Entre la chacra y la olla: Cultura, economía política y las vendedoras de mercado en el Perú*. Lima: Instituto de Estudios Peruanos (IEP).

Buie, S. (1996), Market as mandala: The erotic space of commerce. *Organization*, 3(2), 225–232.

Caravedo, B. (2013), Transformación cultural y cambio de sentido. In: Schwalb, M. & Sanborn, C. (Eds.), *Comida chatarra, estado y Mercado* (pp. 215–222). Lima: Universidad del Pacifico.

Collyns, D., Parkin Daniels, J., Phillips, D. & Agren, D. (2020), 'Hubs of infection': How covid-19 spread through Latin America's markets. *The Guardian*, 17 May 2020 [online]. Available from: https://www.theguardian.com/world/2020/may/17/coronavirus-latin-america-markets-mexico-brazil-peru [Accessed 13 March 2021].

Consorcio Agroecológico Peruano (2020), A la opinion pública. Comunicado de la mesa de rabajo de Políticas para la Promoción de Entornos Alimentarios Saludables. *Consorcio Agroecologico Peruano*, 12 May 2020 [online]. Available from: https://drive. google.com/file/d/1nhczXGpkFSkZVg5KKZ8wg_6xTiGSoP-R/view?fbclid=IwAR1pt-DIDYyvAmvVBLtoDJMIz-80VOJzS7UcNyctWhcRwMQ6YNvAC9vh2BAw [Accessed 13 March 2021].

De Certeau, M., Giard, L. & Mayol, P. (1998), *The practice of everyday life. Volume 2: Living and cooking*. Minneapolis: University of Minnesota Press.

De La Pradelle, M. (1995), Market exchange and the social construction of a public space. *French Cultural Studies*, 6, 359–371.

Díaz-Albertini, J. (2016), *El feudo, la comarca y la feria. La privatización del espacio público en Lima*. Lima: Universidad de Lima.

FAO (2016), "*Los mercados de abastos siguen siendo la principal forma de provisión de alimentos para la población peruana*". 11 February 2016 [online]. Available from: http://www.fao.org/peru/noticias/detail-events/en/c/382956/ [Accessed 13 March 2021].

Filgueiras, B. (2009), *Miradas sobre el comercio popular en la ciudad de Lima. Revista Argumentos*, 3, 20.

Fincher, R. (2017), *Pushing diversity beyond recognition*. Working paper. Melbourne: School of Geography, University of Melbourne.

Fincher, R. & Iveson, K. (2008), *Planning and Diversity in the City: Redistribution, Recognition and Encounter*. Basingstoke: Palgrave Macmillan.

Fraser, N. (2000), Rethinking recognition. *New Left Review*, 3, 107–120.

Friedrich Ebert Peru Foundation (2020), *Webinar: Los mercados de abasto dialogan*. Available from: https://www.facebook.com/686549174727102/videos/732580674173151 [Accessed 13 March 2021].

Higuchi, A. (2015), Characteristics of consumers of organic products and the increase in the supply of these products in metropolitan Lima, Peru. *Apuntes*, 42(77), 57–89.

Huaita-Alfaro, A.M. (2019), *Encounters at Lima's Inner-City Marketplaces: Negotiating Fragmentation and Common Living in Urban Peru*. PhD Thesis [online]. Available from: https://discovery.ucl.ac.uk/id/eprint/10070259/

Huaita-Alfaro, A.M. (2020), Mercados para la nueva convivencia. *La República*, 30 June [online]. Available from: https://larepublica.pe/opinion/2020/06/30/coronavirus-peru-cuarentena-mercados-para-la-nueva-convivencia-por-ana-maria-huaita-alfaro/ [Accessed 13 March 2021].

INEI (2014), *Situación del mercado laboral en Lima Metropolitana. Informe técnico No. 4—Abril 2014*. Lima: Instituto Nacional de Estadística e Informática [online] Available from: http://www.inei.gob.pe/media/MenuRecursivo/boletines/02-empleo-ene-feb-mar-2014.pdf [Accessed 13 March 2021].

INEI (2016), *Censo Nacional de Mercados de abastos 2016. Resultados a Nivel Nacional Instituto Nacional de Estadística e Informática*. Lima: Instituto Nacional de Estadística e Informática [online] Available from: https://www.inei.gob.pe/media/MenuRecursivo/boletines/confetencia_censo_de_mercados_de_abastos.pdf [Accessed 13 March 2021].

Joseph, J., Castellanos, T., Pereyra, O. & Aliaga, L. (2008), Lima, "Jardín de los Senderos que se Bifurcan": Segregación e Integración. In Portes, A., Roberts, B. & Grimson, A. (Eds.) (*iudades Latinoamericanas): Un Análisis Comparativo en el Umbral del Nuevo Siglo*. Mexico: Univesidad Autónoma de Zacatecas, pp. 303–367.

Koch, R. & Latham, A. (2012), Rethinking urban public space: Accounts from a junction in West London. *Transactions of the Institute of British Geographers*, 37, 515–529.

Lima Como Vamos (2020), *Webinar: Mercados y la calle como espacio de trabajo.* Available from: https://www.facebook.com/watch/live/?v=765286940713450&ref=watch_permalink [Accessed 13 March 2021].

Ludeña, W. (2010), *Lima. Reestructuración económica y trasformaciones urbanas. Periodo 1990-2005.* Cuadernos Arquitectura y Ciudad, No. 13. Lima: Pontificia Universidad Católica del Perú, Departamento de Arquitectura.

Medina, F.X. (2013), *Mercados y espacio público: transformación y renegociación de nuevas demandas urbanas. Análisis comparativo de casos (Barcelona, Budapest, Buenos Aires).* Zainak Cuadernos de Antropología-Etnografía. 36, pp. 183–200.

Medina, F.X. & Alvarez, M. (2009), El lugar por donde pasa la vida… Los mercados y las demandas urbanas contemporáneas: Barcelona y Buenos Aires. In: Medina, F. X., Ávila, R. & De Garine, I. [coord.] *Alimentación, Imaginarios y fronteras culturales. Ensayos en honor a Helen Macbeth.* Guadalajara: Universidad de Guadalajara, pp. 183–202.

Ministry of Economy and Finance of Peru (2021), *Programa de Incentivos a la Mejora de la Gestion Municipal del Año 2021. Guía para el cumplimiento de la Meta 6: Regulacion del Funcionamiento de los mercados de abastos para la prevencion y contencion del COVID-19.* Lima: Ministry of Economy and Finance of Peru.

Municipalidad de Miraflores (2007), *Miraflores, 150 años de historia.* Lima: Municipalidad de Miraflores.

Muñoz, H. & Rodríguez, Y. (1999), *Microempresarios: Entre demandas de reconocimiento y dilemas de responsabilidad.* Lima: Instituto de Ética y Desarrollo de la Escuela Superior Antonio Ruiz de Montoya.

Páucar, C. (2020), Los Mercados: Integrados al desarrollo sociocultural de cada distrito. *La República*, 28 May 2020 [online] Available from: https://larepublica.pe/sociedad/2020/05/28/coronavirus-en-peru-los-mercados-integrados-al-desarrollo-sociocultural-de-cada-distrito-covid-19/ [Accessed 13 March 2021].

Peru. Law N° 26569 (1996), *Establecen mecanismos aplicables a la transferencia de puestos y demás establecimientos y/o servicios de los mercados públicos de propiedad de los municipios.* Lima: Congreso de la República del Perú.

Peters, P. & Skop, E. (2007), Socio-spatial segregation in metropolitan Lima, Peru. *Journal of Latin American Geography*, 6(1), 149–171.

Pottie-Sherman, Y. (2011), *Markets and diversity: An overview.* Max Planck Institute for the study of religious and ethnic diversity. MMG Working Paper 11–03. Göttingen: Max Planck Institute for the Study of Religious and Ethnic Diversity.

Protzel de Amat, J. (2011), *Lima Imaginada.* Lima: Universidad de Lima.

Santandreu, A., Huaita-Alfaro, A.M. & Ortega, C. (2021), Los mercados de abasto como centralidades para la resiliencia alimentaria de la ciudad. In: Ocupa Tu Calle (Ed.) *FIIU5. Resiliencia Urbana-Ttomo I: Artículos y Fotoensayos*, pp. 60–69. Lima: Ocupa tu Calle.

Simone, A.M. (2014), *Jakarta: Drawing the city near.* Minneapolis: University of Minnesota Press.

Tello, H. & Narrea, F. (2014), La ciudad y sus mercados: socialidad y cultura en el espacio público. *Peru Debate*, Año 2, 1(1), 40–49.

Vega Centeno, P. (2014), Tenemos más problemas de miedo a la ciudad *La Mula—Un Proyecto de Ciudad para Lima.* 29 September 2014 [online] Available from: https://proyectociudad.lamula.pe/2014/09/29/conversando-con-pablo-vega-centeno/proyectociudad [Accessed 13 March 2021].

Villar, P. (2020), Produce: "Vamos a seguir impulsando los parques industriales y mercados de abasto". Interview to Jose Luis Chicoma, Minister of Production. *El Comercio*, 26 November 2020. Available from: https://elcomercio.pe/economia/peru/diversificacion-productiva-produce-vamos-a-seguir-impulsando-los-parques-industriales-y-mercados-de-abasto-pesca-noticia/?ref=ecr [Accessed 13 March 2021].

Watson, S. (2009), The magic of the marketplace: Sociality in a neglected public space. *Urban Studies*, 46(8), 1577–1591.

11 Markets and belonging

Untangling myths of urban versus small-town life

Sophie Watson and Markus Breines

Introductions

Markets are important social spaces in many cities, towns and villages across the world. While street markets in different locations share some basic qualities in terms of the activities that take place there, the variations between markets have rarely been acknowledged. However, the physical appearance and the institutional frameworks not only differ between countries and regions but also the ways in which traders work in markets reflect different notions of belonging to markets and places. Taking into consideration the variety of street markets and the (im)mobile practices of traders, this chapter examines how social ties to markets are differently shaped in urban and small-town settings in the UK context.

In urban sociology there has been a traditional distinction between the notions of *Gemeinschaft* and *Gesellschaft*, which were typically translated as community and society. These terms have their origin in the writings of German sociologist Ferdinand Tönnies (1887) to designate two dichotomous forms of social ties. In his use of the terms, community (Gemeinschaft) was associated with pre-industrial and predominantly rural living and villages where there were strong social connections and interactions, and shared values and beliefs. The notion of society (Gesellschaft) described looser indirect interactions and connections, more formal values and beliefs. Such notions were deployed to characterise the (imagined) differences between the country and the city. They were further amplified in a plethora of writings where the city has been figured as a site of anomie (Durkheim, 1984), of anonymity (Tonkiss, 2005), of lack of rootedness (Sennett, 2005), of fleeting encounters of the urban flâneur (Benjamin, 1999), of placelessness (Relph, 1976), of a lack of place attachment and so on.

Clearly, these notions have been challenged and unpacked across many studies over the decades. However, a consistent trend suggested by research has pointed to a loosening of urban community ties, as the older suburbs and terrace housing made way for high-rise flats in the 1960s (Wilmott & Young, 1969), where problems of poor infrastructure and lack of investment were a further contributory factor to the sense of alienation and diminished sense of community life that many of the former residents of the old terraces felt. A further development in the last 30 years or so was the mass production of private house and flat developments, which were built for households and individuals working at a distance

DOI: 10.4324/9781003197058-11

from home, often in the central parts of the city, leaving the suburbs early in the morning and returning late at night, which mitigated against the building of strong community ties.

Currently, there are diverse bodies of literature that consider the formation of social relations in urban and rural spaces and which challenge dichotomies of urban/rural as the lines between the two become increasingly blurry, as urban processes permeate non-urban spaces (Andersson, 2009; Lichter & Brown, 2011) and vice versa (Sandstedt & Westin, 2015), particularly through digital processes and new patterns of mobility (Perriam & Carter, 2021). Like Lichter and Brown (2011), we suggest that the distinctions and boundaries between urban and rural communities are often more symbolic than real. There are parallels here between the distinctions drawn between space and place—a dichotomy critiqued by geographers for years (Massey, 1994, 2009). Nevertheless, cities are typically seen as more subject to processes of globalisation and associated rapid changes than more rural areas and small towns.

These socioeconomic transformations have been seen where global processes disembedded the city, and its composite parts, from its rootedness in the local (Sassen, 2001). In this conception, cities are subject to complex patterns of mobility, while small towns tend to have a more stable population (Bell & Jayne, 2006). Thus, city spaces are seen as more transient, disconnected and less embedded in local communities whose members have a diminished sense of attachment to place.

Although this attention to the importance of globalisation in defining cities has been challenged by studies that have reasserted the importance of the local and interrogated the nature of attachment to place and territory, Lewicka (2011) and other environmental psychologists continue to emphasise place attachment as an important part of human existence, while geographers and urbanists have explored the complex meanings of belonging and the role of place and place-making in producing a sense of attachment to the materiality of place (Raffaetà & Duff, 2013). Attachment to place is constituted, we suggest, in an assemblage of social, material, cultural and affective practices that are complex to disentangle. The sense of belonging in a fast-changing world, or attachment to place, may indeed be disrupted by wider social and global processes, which may be more evident and active in city spaces. But (lack of) attachment to place, a sense of local allegiance, may not be easy to read off the surface of a territory, which will change and emerge from "the relationality that is its primary constituent feature" (Raffaetà & Duff, 2013: 341). Thus, migrants may connect across space to family and friends in a complex network of local/global mobilities and connections (Bridge & Watson, 2010: 162).

These different imaginaries and constructions constituted the framework for our explorations on two markets; one located in an urban area (Haverfield), one in a more rural location (Golding; names anonymised). Through the fieldwork on two markets in England, we were keen to unpack the differences between them in relation to notions of the 'local' and attachment to place. Haverfield market is based in an inner-city area with a majority of low-income households, which has become increasingly ethnically diverse over the last few decades with migrants from across the globe. In recent years, it has been subject to processes of gentrification as young professionals have moved further from the city centre in search of affordable housing. A first glance at the market revealed a seemingly more mobile population and more apparent features of a globalised space, such as a larger non-white population among both the shoppers and the traders, as well as the trade of goods whose provenance extended beyond the United Kingdom. Haverfield market is open six days a week. It has been established in the area for many decades.

In contrast, initial impressions of the Golding market revealed a more typical country town homogenous population and a presence of stalls selling goods that appeared to originate in the United Kingdom. The market at Golding on the south coast is only open on Wednesdays. Traders come from different places along the coast to set up their stalls (usually on the same site each week). There is limited ethnic diversity as the traders are predominantly white British, a composition of traders very different from those in Haverfield, reflecting in part the social and ethnic diversity of the wider population in each place. Golding market has had a shorter history in its current location (about 20 years) than Haverfield. Its size is also smaller with about 20 stalls compared to Haverfield with more than 100 stalls.

We regularly visited the two markets from the autumn of 2019 to spring 2020 when the COVID-19 pandemic restricted our research and spent time observing and speaking to traders while they were working. This was supplemented with semi-structured interviews with market managers and traders. After lockdowns in the United Kingdom began, we continued our research through interviews on FaceTime and Zoom. Interviews covered life histories, trading practices, mobilities, choice of commodities, social relations, experiences of everyday life and work patterns. Through this research, it emerged that there was a complex relationship between locations and market traders' senses of belonging, which did not map easily onto the urban/rural divide. In this chapter, we examine how different patterns of mobility among the traders in these markets reflect very different belongings to place and argue that an exploration of the mobile practices of market traders can untangle some of the myths of urban versus rural life where city spaces are often conceived as more fluid, less fixed, with weaker social ties, when compared to more rural or small-town spaces.

A sense of place in an urban market

Contrary to the dominant imagination of contemporary cities as transitory spaces, where retail outlets, restaurants and personal services come and go as fashions wax and wane, the strength and longevity of Haverfield market were revealed in several ways. The most significant factor was that many of the traders in the market had worked there for many years, a point made to us by Samir the market manager, which was confirmed in our interviews with traders. As a result, these traders described a network of connections to the locality, having worked on the market for decades. Many of these traders had been brought up and gone to school there and knew many of the local community, as well as other traders, something we witnessed time and again as they hailed people walking by. Their mobility patterns were thus very limited, with some having barely left the area—even to travel into the city centre.

Haverfield market is open throughout the week, and many of the traders work their stalls each day, typically on the same site with the same infrastructure. One vegetable and fruit seller, Fred, provided a rich picture of his local connections, which was replicated across the market. His story is emblematic of many other traders. Fred had spent his whole life working in this same business after taking it over from his father who had started trading in the market just after the Second World War. Fred was born in a local hospital and had grown up in the area. As a child, he regularly frequented the market with his parents and their friends where much of his social life was enacted:

> I used to come with my mum and dad, sit on the stall and my mates were all there. My mum and dad's friends who used to work in the fish shop and we used to play together, and we was just pals, it was a community.

His father owned the stall before him, so from an early age, he had been involved in the everyday labour of the market. Similarly, his father remained working on the stall with Fred until his health no longer permitted it:

> I was really small, my dad used to work and I used to push my little barrow around, used to be behind the butchers, and they used to have nine boxes, that's all we had on the stall. Nine little boxes, and I used to put the scale and the grass in it, and I used to push it round and come out the back of the butcher's and I pushed it down the back where there used to be a pie factory, where these houses are, at the top of the High, there used to be a pie factory called Gillard's Pies.

While his childhood contribution to his father's business illustrates his close relationship to the area, Fred's long-standing relations on the market had transformed over time. The ethnic diversity of the area—rather than causing dissonance and resentment, as some have argued as characteristic of major cities (Wells & Watson, 2005)—was something that Fred also celebrates as part of the community:

> I serve a lovely family, Asian family, up there one the left. Been friends with them for 30 years, absolutely lovely family. They come to me, bring, in Diwali he brings me some chicken, brings me rice, brings me a Christmas card. It's so nice. And Mr Ali, one of my fellow traders, he always buys me a packet of biscuits for Christmas, always wishes me a happy Christmas. He comes and has a cup of tea with me every … comes to my stall at 9 o'clock every day and makes his tea, has a cup of tea. Lovely man. Lovely, served all his family, served his cousins, all of them they come to us.

This description of attachment to the area and the market, in particular, was not restricted to long-standing white working trading families in the area but was evident in the more recent migrant families to the area. Ahmed, a trader of sheets and bed linen, gave the following explanation of how he came to work on the market:

> It was local, I was living in Haverfield at the time and I feel like I was born in Haverfield, I went to school in Haverfield, I told my kids I'll probably die in Haverfield, so it's just, I used to remember going to the market as a child with my mother, and she used to hold my hand so tight because there were so much public that you can't even move. So, you know, literally you're like holding your mum's hand so tight so you don't get lost. And I remember that market vividly, it was just so vibrant, we used to enjoy our days out going to the market, a nice family atmosphere, it was just lovely, it was just great.

There was a large proportion of traders who originated in other countries and had moved to the United Kingdom like Ahmed. Many of them also expressed a strong sense of belonging and attachment to the market and the area. Women traders were part of this narrative also. Two female traders who sold hot Turkish breads stuffed with cheese, meat and vegetable from a van, had come with their families from Turkey to the United Kingdom in the early 2000s and both of them had given birth to their children locally. They lived within walking distance from the market and had used it regularly in the years before setting up their own stall seven years ago. Despite having a different background from traders like Fred, they had built up a regular customer base and worked closely with a café owner in front of their stall, which was in the same spot every day. Others traded on specific days, which were usually consistent so that customers knew when to find them, such as the Afro-Caribbean seller of herbal remedies who only works on the later days of the week. Their rhythm of working in their own neighbour-hood and localised mobilities contributed to making them an important part of the market but also to embedding them in the social relations that shaped a sense of local community.

Because of all these traders' history in the area, they articulated a sense of belonging, which derived from close ties to the neighbourhood and the people in the community. For them, the market was important not only as a place of work and income but also to be part of the community was very significant to them and their families, many of whom worked on the stalls also. Like traders in other studies (Watson, 2009), those interviewed in Haverfield played an important role in the community looking after people who were vulnerable or lonely, such as some older people, for example, in the form of showing an interest in their lives and families, and in some cases even helping them out in practical ways, such as dropping off food for them. Traders here see the market as a community. As Mo, the Afro-Caribbean perfume seller said,

> I think markets are very important because it brings more of a local, com-munity out. I think with shops, shops bring too much professionalism to the role, while with markets it's more of a place people can gather, socialise, and it's more of a community sort of thing, where it's not just trading, but you're

trading amongst the community that you all get to know each other and be all ... like a lot of people, a lot of customers they come, they stop, they talk, and that's what I like about the market community, you know people and look after them.

This 'caring' role is an often-ignored part of the market life in local communities that has been mobilised as an argument for greater support for markets at the governmental level.

In addition to his emotional attachment to the market, Fred was very conscientious about his produce and took great care to adapt to social and demographic shifts in the local area:

All the English produce obviously comes through England, from the farms and different distribution companies, so knit into the market. But things have changed dramatically because English produce now, English apples, the Asian community don't really like English apples so much. They like very highly coloured apples, Pink Lady, Royal Gala, red apples. Don't like the old-fashioned Russets and Coxes. That's how your business has changed, you understand that. They like the sweeter flavours of apples.

Rather than global shifts disrupting his strong sense of belonging and pride in the history of his family's relation to the market, he had modified his business and practices to accommodate the impacts this had on the market.

Mobile lives in a non-urban market

There are fundamental differences between urban and rural markets. These differences have emerged historically. According to Joe Harrison (chief executive of the National Market Traders Federation), in the 1970s, the majority of markets, both rural and urban, took place on one day a week. This 'market day' was popular and drew in large parts of the local population. As a result, market trading was economically successful as far as the traders were concerned, who on the following day headed off to another town or city. In each new venue, there was a new population who shopped in the market, and who like their counterparts in other markets fulfilled many of their food and other consumption needs there. This has changed over time as some local authorities, particularly in cities, have pushed for daily markets in their areas to increase the rental income from the market stalls occupied permanently. However, if a market is open every day, there is not always a corresponding increase in customers or in income accrued to the trader. Despite the increase of markets running on a daily basis (like Haverfield), the weekly market—typically described as the market day—is still common on the south coast of England (like Golding). This pattern has different implications for the ties between traders and customers compared to places where they take place more frequently.

Golding market is open on Wednesdays; on the other days of the week, the traders go to other markets along the south coast from Rye in East Sussex to

Portsmouth and beyond. Some of them would trade in their hometowns once a week but were highly mobile the rest of the week. This mobility meant that they kept their goods or produce stored in their vans overnight to facilitate their continuous movements between markets. Few of them covered the same markets as one another because each of them had preferred markets and routes, which meant that it was not only the connections to the customers that were limited but also to other traders. They would, however, stick to markets they found to be working for them, and there was a group that had worked on Golding market for several years. The market manager, Clare, explained that

> the belt and braces I call them; they've been doing it for years, they know their product, they won't change, but actually they're a good backbone to the market.

Their regular presence and successful businesses generated consistency to the market and helped build up social relations between customers and traders. But these traders were in the minority.

Overall, the frequent movements of the traders made the sense of community less prominent than in Haverfield; few of the traders emphasised their sense of attachment to the market or an extensive knowledge of the local community. John, a man selling cheese, worked every day except Monday and Thursday in five different markets. He followed the same route on a weekly basis and interacted with large numbers of customers and traders on a weekly basis. He did not live in Golding, and this market was just one out of the markets that he spent a day in, which meant that his ties to Golding were less strong than the traders in Haverfield who had developed wider connections to the local community through family, school or social connections over many years. John and several other traders in Golding were continuously adapting their businesses and sometimes changed the locations where they were trading, which can be seen as a result of their loose social ties to places as well as practices that would make their social ties less place-based.

Another difference from the Haverfield market was that there was also the presence of traders who aimed at more upscale customers, including tourists to the town. Some of these were individual traders and others represented companies that had a presence in different locations—for example, at 'foodie events' in the region. Clare explained,

> So, they'll be in Golding on a Wednesday, but they'd also be somewhere else. So, it's a bit like the Marks & Spencer's of the market world! They've got four or five different set-ups and they do lots of events, they provide to commercial premises, so they're quite a modern one.

These ways of trading, which often involved hired staff, are in contrast to traditional ways of working. These employees who work for companies and move between markets, as well as events in other locations, have fewer opportunities to establish relationships with customers than traders who work on the same market every day.

Even though the traders did not necessarily have strong relationships with Golding, the customers were largely local. Clare explained this by referring to a market research study conducted by the council which found that 86% of the people passing through the town centre where the market is located were found to be local residents. Notwithstanding the high loyalty rate, Clare was eager to make the market an attraction that would bring in people from a bit further afield:

> Once you've started to capture them and they get into the habit, they keep coming, which is a positive. So, it's about dragging people and drawing people in.

Although the traders were less invested in the market than the traders in Haverfield, they took on an important role in facilitating a social space in the city centre. By trading on a weekly basis in Golding, the traders contributed to gathering the local community, as well as a small number of people from elsewhere, around the market. The traders played a central role by creating a meeting place and regular event for the local community, even if their own relations to those communities were less central in their work, and they did not gain a strong sense of being part of the community themselves.

Some of the longer-term traders in Golding who trade at that site on the same day each week from year to year were known and recognised locally. Yet, none of the traders interviewed described a strong historical sense of belonging and attachment to the local community. The fact that the traders were more mobile than the traders in Haverfield appeared to impact on their sense of belonging and their sense of being part of the community very differently from the traders who lived proximate to the site and returned to the same market on a daily basis. As a result of their regular rhythms of mobilities, they had developed connections to other people and places which made their social ties more dispersed and less place-oriented than in Haverfield. It was thus clear that the imaginary of markets in smaller towns as 'local' and more embedded and community-based when compared to urban markets which are intricately embedded in processes of globalisation with seemingly more transient populations does not necessarily reflect the realities of the relations between traders and local communities.

Conclusion

In conclusion, the traders in Haverfield were more attached to the market they were working in and had a stronger sense of community than the traders who worked in the smaller markets on the south coast. There appeared to be several reasons for this. In urban Haverfield, the intersection of strong social relations between traders and with shoppers constituted over time, even where people were connected elsewhere also, the stability of work practices in one place, the daily and repetitive material practices of setting up and dismantling the stall in one site, among other factors, worked together to create a strong sense of attachment to the local place and strong community ties. The mobility patterns of the traders in the south coast market meant engaging in daily journeys to markets beyond their home place. Many of the traders lived in another town and had limited local knowledge or connection. As such, few described a sense of attachment to the market but talked instead more about their product than the local community.

This small study of two markets thus complicates the narrative of inner cities as likely to be marked by weaker social ties and smaller towns and villages to be characterised by stronger social connections. A sense of place and belonging derives from social, historical and economic relations that cannot be easily read of imagined urban/rural differences. This is interesting also in the context of the institutional framework of markets. The National Association of British Markets Authorities has run a recent campaign to support and enhance the attraction of markets called: *Love Your Local Market*. Such a discourse implies a notion of localness which frames markets as 'in place'. Yet as we have shown, they are temporary or less temporary events to which the traders may have different attachments.

Nevertheless, it is important to recognise that in each site, despite the different mobility practices of the traders and their different relationships to the markets they were working in, they contributed to building communities through their presence and activities in different ways. Markets are important sites of community building regardless of the frequency of traders and have been found to be central in this and that (Watson, 2006). They offer a public space which is open and inclusive, not highly regulated in terms of social encounters and which offers serendipitous opportunities for surprise and delight in ways that are not replicated to the same extent in supermarkets or shopping centres (Watson, 2006). At the same time, by demonstrating that market traders have different relationships to markets, this chapter has highlighted the spatially differentiated ways in which traders contribute to shape communities.

The notions of *Gemeinschaft* and *Gesellschaft*, though less central to urban studies than a century ago, retain a certain force in urban imaginaries. So too notions of the local and of place have been increasingly problematised as this chapter aims also to do. Urban and rural communities have become increasingly complex assembling complicated relations of people and things. Nevertheless, urban/rural distinctions remain forceful in many different arenas. By providing findings from this research that challenge these narratives, this chapter has demonstrated that the contours of social relations and community in cities or small towns cannot easily be read off their apparent spatial configuration or presence. Through the investigation of the mobilities and practices of building community and a sense of belonging in two different markets, we demonstrate that the rural-urban distinction needs to be constantly unpacked and explored in order to understand the specificities of place.

References

Andersson, K. (2009), *Beyond the Rural-Urban Divide: Cross Continental Perspectives on the Differentiated Countryside and Its Regulation*. Bingley: Emerald Jai.

Bell, D. & Jayne, M. (2006), *Small Cities*. London: Routledge.

Benjamin, W. (1999), *The Arcades Project*. Cambridge, MA: Harvard University Press.

Bridge, G. & Watson, S. (2010), Reflections on mobilities. In: Bridge, G. & Watson, S. (Eds.), *The New Blackwell Companion to the City* (pp. 157–169). Oxford: Blackwell.

Durkheim, E. (1984[1893]), *The Division of Labour in Society*. London: Palgrave Macmillan.

Lewicka, M. (2011), Place attachment: How far have we come in the last 40 years? *Journal of Environmental Psychology*, 31, 207–230.

Lichter, D. & Brown, D. (2011), Rural America in an urban society: Changing spatial and social boundaries. *Annual Review of Sociology*, 37(1), 565–592.

Massey, D. (1994), *Space, Place, and Gender*. Minneapolis: University of Minnesota Press.

Massey, D. (2009), The possibilities of a politics of place beyond place? *Scottish Geographical Journal*, 125(3–4), 401–420.

Perriam, J. & Carter, S. (2021), *Understanding Digital Societies*. London: Sage.

Raffaetà, R. & Duff, C. (2013), Putting belonging into place: Place experience and sense of belonging among Ecuadorian migrants in an Italian Alpine region. *City and Society*, 25(3), 328–347.

Relph, E. (1976), *Place and Placelessness*. London: Pion.

Sandstedt, E. & Westin, S. (2015), Beyond gemeinschaft and gesellschaft: Cohousing life in contemporary Sweden. *Housing, Theory and Society*, 32(2), 131–150.

Sassen, S. (2001), *The Global City: New York, London, Tokyo*. Princeton, NJ: Princeton University Press.

Sennett, R. (2005), Capitalism and the city: Globalization, flexibility, and indifference. In: Kazepov, Y. (Ed.), *Cities of Europe: Changing Contexts, Local Arrangements, and the Challenge to Urban Cohesion* (Chapter 5), pp. 109–122. Oxford: Blackwell.

Tonkis, F. (2005), *Space, the City and Social Theory: Social Relations and Urban Forms*. Oxford: Polity.

Tönnies, F. (1887), *Gemeinschaft und Gesellschaft*. Leipzig: Fues's Verlag.

Watson, S. (2006), *City Publics: The (Dis)enchantments of Urban Encounters*. London: Routledge.

Watson, S. (2009), The magic of the market place: Sociality in a neglected public space *Urban Studies*, 46(8), 1577–1591.

Wells, K. & Watson, S. (2005), A politics of resentment: Shopkeepers in a London neighbourhood. *Ethnic and Racial Studies*, 28(2), 261–277.

Wilmott, P. & Young, M. (1969), *Family and Kinship in East London*. London: Pelican.

12 The role of mobility and transnationality for local marketplaces

Joanna Menet and Janine Dahinden

Introduction

> I would say the market on the main square is an important local tradition, one can meet people, go shopping, have a coffee on the nearby terrace; many inhabitants of this city are quite attached to it.
>
> (Interview with public officer, 28 November 2019)

Marketplaces have been described as flexible spatial and temporal assemblages that can contribute to inclusive public spaces (Schappo & Van Melik, 2017). Markets are seen as local events, which allow meaningful encounters between inhabitants (Watson, 2009). Similarly, fresh food markets and farmers' markets are often presented as places where local consumers buy directly from farmers (Alkon & McCullen, 2011).

This chapter explores how the local character of markets is produced by different market participants, such as customers, farmers/vendors and local authorities. We delve into the different meanings of 'local' to further understand how the local is performed and to demonstrate the consequences of vendors' (marketing) strategies, be they deliberate or unintentional. We show that the local is entwined with mobility and transnationality, both elements that are invisible in common representations. Furthermore, we argue that markets can be analysed as expressions of national forms of identification.

This chapter is based on ethnographic fieldwork on two fresh food markets in Switzerland as part of the research project *Moving Marketplaces: Following the Everyday Production of Inclusive Public Space*. Starting in December 2019 and taking place for seven months, the researchers participated in fresh food markets as observers, customers and vendors. During their working days on the markets, we followed vendors and visited some of them on their farms or their further selling points. In addition to participant observation, including many informal interviews with market actors such as vendors and customers, we conducted 13 in-depth interviews with six producing and three non-producing market vendors, two market managers and two city authorities. As part of the research was conducted during the COVID-19 pandemic, and some markets were cancelled for several weeks, we adjusted our methods, and included phone interviews (MMP Team, 2020).

DOI: 10.4324/9781003197058-12

The two cases we studied are located in two linguistic regions of Switzerland: a weekly fresh food market in the country's biggest city of Zurich, and a fresh food market taking place twice a week in the smaller town of Neuchâtel, which is close to the French border. Both markets are renowned for their qualitatively high offer of local food and cater to rather affluent customers. A peculiarity of Swiss markets is that food markets are strictly separated from non-food markets and that there are no market halls with permanent stalls. Instead, the municipalities assign particular places within the city for the markets—places that are used for other events or simply stay empty during non-market days.

We adopt a mobility approach (Cresswell, 2010; Sheller & Urry, 2006) and a transnational perspective (Dahinden, 2017; Glick Schiller et al., 1992) to grasp the significance and construction of the markets' local character. Embracing a mobility perspective is a means to overcome the 'sedentary bias' still widespread in social sciences (Urry, 2007)—namely, a position which normalises stillness and immobility. Instead of assuming rooted and bounded communities, a mobility lens challenges such static notions of society. It opens up possibilities to investigate the various movements of vendors as well as their products, be this on small or broader spatial scales. A transnational perspective stands for endorsing a particular, alternative stance on social science issues, to overcome methodological nationalism and to go beyond the 'national container' (Wimmer & Schiller, 2002). As we will show, these lenses allow for unpacking multiple dynamics which produce the local character of markets.

As we observed, the local is imbued with different meanings, sometimes relating to the vendors but often also to the products sold on markets. Literature on local food systems and food localism demonstrates the complex and contested meanings of the local when it comes to food (Forney, 2016). Indeed, "local food means different things to different people in different contexts" (Eriksen, 2013: 49). In this chapter, we analyse these different meanings and their links to mobility and transnationality.

Our conclusion shows that marketplaces can be understood as a particular local expression of global identity politics. Markets are local only insofar as vendors apply many strategies to bind their clients in proximity—but their locality is also conditioned by their embeddedness in global identity politics where mobility, transnationality and nationalism play a crucial role in the economic success of markets.

Relational proximity and the mobility of vendors

> And for me, it is really about the short distances, the proximity of the citizen with his or her farmer.
>
> (Interview with city councillor, 23 April 2020)

Food markets are often imagined as facilitating direct contact between producers and consumers, a relationship presented as "immediate, personal and enacted in shared space" (Hinrichs, 2000: 295). Yet, dwelling more deeply within this assumed natural locality of markets allows us to unpack interesting aspects regarding the ways in which market actors produce proximity. Relational proximity is often presented as having particular qualities (Alkon & McCullen, 2011), which ideally embody mutual, long-established familiarity and trust. As studies on local food systems have argued, such a face-to-face interaction between consumers and growers is opposed to large-scale, industrialised systems of food production and distribution (Eriksen, 2013; Hinrichts, 2003), and this is an important facet upon which outdoor markets capitalise.

As in the previous quote, the fresh food markets in our case studies were not represented as places of *casual* encounters between vendors and customers. Instead, the city councillor draws on a (romanticised) image of proximity in terms of market relations between producers and consumers. Indeed, customers showed great solidarity and fidelity when markets had to close down due to restrictions during the COVID-19 pandemic. One vendor told us, "People were really loyal (…) I received many messages like this, people really wanted to support us, to continue buying our produce" (interview with vendor, 18 May 2020). In line with customers' wishes to support regional producers, the value of solidarity becomes enacted in relationships between farmers and local consumers (Forney, 2016).

Market vendors were thus aware of and enacted the image of relational proximity through multiple strategies. A young vendor told us that in her experience, it is important to always "have the same face behind the stall", to build up a relationship with customers. Her employer, an old-aged farmer, would always be present during market hours and available for a brief chat with clients, an activity she believes encourages customers to return. Similarly, a successful vendor with several employees told us that he always asks his selling personnel to smile and talk with their customers because "if people come to the market, it's to have someone to talk to".

Establishing personal connections with customers is a widespread practice of farmers to build up regular customers. To do so, vendors chat about the weather, different farming methods or gardening techniques, and one vendor told us he even invites his regular customers to drink a coffee from time to time, as "this is part of it". We also observed vendors offering their clients some parsley, coriander or chives for free, and when asked, several explained that it was to emphasise their difference from anonymous supermarkets.

However, to become recognised by their clients, stallholders insist on their exact and consistent emplacement. Changes in the spatial allocation of the stalls on the market—for example, due to construction work, temporary utilisation of the place for another event, or during the COVID-19 pandemic—are met with protests. Vendors told us that they need to be precisely at the same spot, otherwise their clients would not find them. One vendor stated that he is the third generation selling on a particular spot, thus stressing a historically anchored link to the specific location.

Furthermore, in many markets, vendors post a plaque displaying their family name—often identifiable as local—and the residence address. Here, the institutional framework in the form of the written market regulation stipulates the vendors' proximity strategies: it transmits the idea that the vendors and their products sold at a particular stall originate from the nearby village indicated on the plaque.

However, this drawn image of markets as spaces of relational proximity between local farmers and their customers masks the translocal or transnational networks and mobility practices of most market vendors. The relational aspects of the local are only enabled by vendors' translocal and transnational mobility. While some of the vegetable sellers do indeed grow their own products, all the farmers we talked to also buy a part of the fresh produce they sell on the market (see the following), and to do so, they are mobile.

Some vendors do not grow anything themselves but buy all the produce from farmers or wholesalers to resell them on the market, a fact their customers are often not aware of. This practice is embedded in multiple mobility patterns as well. Some vendors spend the day before a market driving around to collect produce from farmers. Others order from intermediary vendors or drive to neighbouring France once a week to buy fresh produce and import it to Switzerland. A vendor told us that he sometimes travels to Southern Italy to directly meet with a distributor, who is himself in contact with the farmers: "I already went to Sicily four times to meet him. With EasyJet it is simple, you get a ticket for 80 Swiss Francs, and you fly to Sicily" (interview with vendor, 12 June 2020).

As these cases demonstrate, the networks between farmers, wholesale markets and market vendors often span regional and national borders. The produce has already been sold and resold several times before it arrives on the marketplace. As vendors and products circulate, they become only literally localised during the market day. Yet, through the previously described practices, these issues remain in the shadow with only the representation of proximity being produced.

Moreover, vendors usually drive to the marketplace from outside the city. As part of this mobility, vendors move their selling point with them, a highly labour-intense practice. As one vendor put it,

> We have to get up at 4 or 5 am to prepare the crates and then come to the market and build up the stall. Every single time (…) though there are supermarkets with everything you can dream of. And just next to them, twice a week, there is a kind of absurd theatre going on. We build up stalls with planks of wood, we put the crates and then we sell salads. (…) It is really like a kind of theatre from another time that happens in the middle of this ultra-technical society. What a paradox!
> (Interview with farm employee/vendor, 18 April 2020)

When customers arrive at the market, all they see are the already installed stalls offering 'local products' and with the well-known faces behind them. The mobility of the vendors and their products remain invisible as they perform the 'theatre' which produces markets as local. Put another way, the image of local markets as natural encounters between vendors and consumers is built on the idea of proximity, which is the result of vendors' high investment in localisation and mobility practices.

Markets as expressions of banal and affective nationalism

> People (…) really want seasonal vegetables that are from here… that don't come from abroad. Actually, that's the reason for people to come to the market. They want fruit and vegetables that are Swiss, not coming from Italy or abroad.
>
> (Interview with farmer/vendor, 30 March 2020)

As this quote suggests, customers choose food markets when looking for local produce. However, a product being considered as local is the result of all market actors' negotiations and imaginaries. This becomes clear in the following quote by a producer who told us that first she only sold homegrown vegetables, but then she started to enlarge her offer with other Swiss products:

> Even older clients, when they ask "ah, you already have your own tomatoes?" and I say "no, it's not my tomatoes yet, but they are Swiss", they are okay with it and take them. I wouldn't dare to sell them strawberries from Spain, no way. You need to keep this, it has to stay Swiss. Oh, that's important to them.
>
> (Interview with farmer/vendor, 15 June 2020)

As this quote demonstrates, the idea of local food as having been grown in close geographical proximity often intermingles with its national origin. Meanings of a product's locality here intersect with national boundaries and a national imaginary of Swiss products being the 'best'. This is probably a particular phenomenon for food markets, given that there exists a specific link between what is considered to be 'good' when it comes to food. The latter is based on ideas of the freshness and quality of the products, as one vendor put it: "These vegetables just taste differently, fresh and better". Furthermore, the criterion of quality is often related to ecological arguments such as less transport miles or the seasonality of products on the market. Consequently, buying Swiss products rather than imported ones is perceived as more environmentally friendly. Indeed, in Switzerland, the view of Swiss agriculture as being more ecological than in other countries is widespread, as Forney and Häberli (2016) found.

However, as those authors argue, "this belief has not been empirically evidenced" (Forney & Häberli, 2016: 148). Rather, the idea of local food as automatically being better is part of what DuPuis and Goodman (2005) have identified as an 'unreflexive localism' in alternative food movement activism. During our fieldwork, we overheard customers asking for Swiss produce, which illustrates the tendency to estimate that local food is better. Thus, vegetables and fruit which are produced in Switzerland acquire value as such. In other terms, 'local' often means grown within the national borders of Switzerland.

We would therefore go one step further and argue that markets can be analysed as an expression of what has been called *affective nationalism* (Antonsich & Skey, 2017): market customers display an affective and emotional relationship to Swissness when they automatically assume that Swiss products are the best and when vendors put forward the same image. Additionally, marketplaces have a crucial role in what Billig (1995) labelled *banal nationalism*: Swiss products on the markets turn into everyday representations of the nation, which build a shared sense of national belonging linked to a particular (national) territory. Indeed, some market vendors carefully pay attention to communicating their products' Swiss origin by displaying the Swiss cross on the crates. Markets turn here into places where relations between people and products underpin national forms of identification and expression. In a nutshell, Swissness makes the markets work.

As any form of nationalism always goes hand in hand with closure and containment within the 'national container' (Wimmer, 2002; Basch et al., 1994), it does not come as a surprise that simultaneously transnational actors and links in the making of the Swiss, local agricultural products are rendered invisible. In fact, our transnational and mobility perspective helps depict many ways in which this affective and banal nationalism works as a filter producing blind spots for the transnational connections and embeddedness of the markets: most of all, they render invisible migrant labour which is based on transnational mobility.

While customers on the market usually come into contact with Swiss farmers or their Swiss student employees, they rarely meet all those who have also worked to plant and harvest the produce. Analysing farmers' markets in the United States, Alkon and McCullen (2011) observe that the Latino/a farmworkers, who are instrumental to producing the food, seldom sell at farmers' markets. "Indeed, despite consumers' assumptions to the contrary, many market farms rely heavily on non-family labour" (Alkon & McCullen, 2011: 946). In Switzerland, additional non-family labour gradually becomes more important, especially during the harvest season (Arnaut et al., 2015). The Swiss farmers union estimates 30,000 to 35,000 temporary migrant farmworkers being active in the Swiss agricultural sector, recruited mainly from within the European Union, especially from Poland, Romania and Portugal (Jaberg, 2018). Most have employment contracts of three to nine months' duration and work 50 to 55 hours per week. As agriculture is highly labour-intensive, salaries are comparatively low, and "no Swiss wants to do this", as one farmer said during his interview.

The situation of temporary migrant farmworkers is rarely a topic in public media coverage nor in market participants' discourses. Customers usually ignore that temporary migrants perform the hard, low-cost labour on Swiss fields. Although products are locally grown, it is only through the mobility of a non-local workforce that production is even possible. Presenting products as of Swiss origin, thus drawing on the image of local as Swiss, market participants render invisible the (transnational) mobility of the workers necessary to produce 'local' food.

Reterritorialisation of local products from elsewhere

If you want to distinguish yourself [from supermarkets] you need products that stand out from the ordinary. So, I also prepare the olives, the sauces, the mix. That's what customers like, because it's "homemade", so to speak.

(Interview with vendor, 13 June 2020)

Interestingly, market participants reproduce the image of the products sold at Swiss fresh food markets as being local, even though part of the offer is imported from abroad, such as olives in the previous quote. Here, the local acquires another meaning, as it implies specific values, even if defined on transnational grounds. As in the quote made by a woman selling olives, tapenades and vinegar, market participants construct markets as local in terms of the national origin of the products and specific values attached to the products. Thus, even products grown in other countries may be labelled and appear local—though sourced from faraway.

The image of markets as spaces where local Swiss products are sold is upheld, although many fruits and vegetables cannot be cultivated in Switzerland, particularly during the winter season. All the producers to whom we spoke who sell their own 'locally' grown fruit and vegetables also offer imported products. This is regarded as necessary in order to economically survive as direct-vending farmers. As one farmer in his sixties put it, "[Y]ou know, were I to try to sell what my parents brought on the market, it would never work out." Laughing, he told us about the old times when farmers would bring two baskets of potatoes and carrots to the market. He had to diversify his offer, in the same way as all the other farmers and vendors on the market, who also told us that their customers value a large offer. Similarly, a fruit farmer who grows apples, cherries, plums and berries and directly sells all his fresh products on two city markets, told us that his clients ask for 'exotic fruit' such as bananas. Like all the other vendors, he buys them at the large wholesale market on the night before the market, where he then sells about 150 kg of bananas per week.

Besides bananas, avocados and other so-called exotic fruit produced on other continents, Swiss market vendors also sell flowers grown in the Netherlands, apricots and olives grown in France and vegetables such as tomatoes, eggplant, red pepper, lemons and oranges, grown in Italy and Spain. Vendors present these vegetables usually in green crates instead of the original packaging, which remind customers of, for example, the broccoli's real production place outside of Switzerland. Even though vendors inform their customers of the export country when asked directly, they render invisible the non-Swiss origin of many of their products.

Market vendors also contribute to the banal and affective elements of nationalism as part of their business strategies. A vendor, who described himself as someone who sells carefully selected products like fruit and olives of good quality, told us,

> Sometimes it's funny. People don't want anything from Spain because they were made to believe that only Spain treats fruit [chemically]. So a Spanish peach: no, but an Italian peach: yes. Even if it is probably treated in exactly the same ways, or probably even more in France, but "France is the best", then Italy, and then Spain.
>
> (Interview with vendor, 26 May 2020)

The symbolic value attached to supposedly less chemically treated fruit is also visible in the prices. According to the fruit vendor, fruit from France is the most expensive, followed by products grown in Italy and then Spain. As revealed in the interview with the fruit vendor, market vendors have to navigate their customers' demand for food they symbolically value more due to a specific provenance. Some vendors choose to import fruit grown in specific regions in France and also travel to those places "so that, when people ask questions about apricots, I can tell them exactly where they come from, how everything works". To re-inscribe their offer into the calls for qualitatively outstanding products, vendors present theirs as individually chosen, qualitatively outstanding products with a clear origin in a place.

As Forney and Häberli (2016) write in a study on localisation strategies in the Swiss dairy industry, every "food product has an origin, but only in specific contexts does origin become a value" (Forney & Häberli, 2016: 140). Indeed, French-speaking market participants in our research used the notion of *terroir* to refer to the origin of some regional specialty products. This notion encompasses more than the aspect of a specific soil or origin, as it connects the history, the place and the specificities of a product (Barham, 2003; Forney & Häberli, 2016). Market vendors have built on the supposed value of specific origin, using what can be termed a 'reterritorialisation strategy'. The latter is framed in the nation-state logic: products are presented as French, Spanish or Italian. Furthermore, the countries are hierarchised in terms of the assumed quality of products. The nation-state logic serves here to link the idea of *terroir* to places outside of Switzerland.

Hitherto, this reterritorialisation strategy works not only for products, but also for the people selling the product. For example, we observed cheese vendors, who sell cheese made in Italy, using Italian words to attract customers and thus mobilising ethnicity in a context demanding for a specific type of authenticity (Menet, 2020). Similarly, a migrant vendor who imports fruit from central Africa to sell them on the market advertises her fruit as 'exotic' and 'wild', and directly demonstrates to her customers how they should cut the fruit. In performing her knowledge thus, she purposefully renders visible her transnational links. The sedentary element inherent in the nation-state logic (Dahinden, 2012; Ghorashi, 2017) serves, in this case, to transnationally embed a product in its (often national) *terroir* and territory and to transnationalise the meaning of local. Vendors position themselves in this transnational logic by mobilising the representation of authenticity: to do so, in the case of vendors of symbolically higher valued specialty products, they create the idea of authentic (local) products from elsewhere. Through reterritorialisation, vendors actively keep up ideas of locality even if this local is faraway, either by linking it to a territory or through their own national origins.

Conclusion

In this chapter, we set out to explore the different meanings of the local within two Swiss marketplaces and how they are (re)produced by market vendors. We identified three different meanings of 'local' and argue that they are the result of market participants' practices, be they deliberate or unintentional. The first meaning is *relational proximity*: at first sight, markets are local and temporally and spatially confined, as vendors and consumers meet in co-presence. Vendors apply different strategies to create not casual encounters but long-lasting relationships. The local emerges through these tactics which aim at binding customers to particular vendors. However, these local relational binding efforts rely on practices that are anything other than local: instead, this relational proximity is enabled by the mobility and translocal and transnational practices of the vendors and the circulation of their products—practices that often remain hidden from the customers.

Second, products are interpreted as local as long as they are grown within the *Switzerland's national borders*. Local in this sense is about *Swiss* products: the value of Swissness is displayed in everyday encounters and consumers assume that Swiss equals quality. Therefore, markets can be seen as places based upon and reproducing banal nationalism as much as affective aspects of a shared sense of national belonging. Nevertheless, every form of nationalism produces blind spots, here the networks that go beyond the national container, such as migrant farmworkers harvesting Swiss produce.

The third meaning of the local is a *transnationalised* one based on the idea of *terroir* and anchored in the sedentary logic of the nation-state. In this meaning, local turns into 'local from somewhere else'. To re-inscribe their products into the calls for qualitatively outstanding produce, vendors present theirs with a clear national origin while different origins become hierarchised in terms of the quality of the products. This 'reterritorialisation strategy' works for products as much as for vendors who mobilise their transnational links in order to highlight the authenticity of their products of a particular origin and their knowledge about the history of the place and product.

These results deserve further discussion, as they give insights into broader social processes. We argue that markets play a role in global identity politics in at least two different ways. First, based on an analysis of the 'local food systems' literature, Eriksen (2013: 48) suggests a taxonomy of local food: "geographical proximity, relations of proximity, and values of proximity". These three domains crosscut with our results and are of crucial importance to markets as local as we have shown. We argue, however, that this literature omits mobility and transnational aspects, with the risk of reproducing a sedentary bias. The mobility of vendors and concomitant circulation of products are fundamental for producing the local character of markets. The markets in our study can be understood as a hinge between the local and transnational, as they are situated at the intersection of different global cultural and economic flows (Appadurai, 1990). Markets are embedded in and simultaneously reproduce a global identity politics where nationalism is one of the crucial motors. By their production of locality, markets underpin national forms of identification and expression, and these issues are important facets of the economic functioning of markets.

Second, there is abundant literature depicting the crucial role food plays for migrants (Morasso & Zittoun, 2014). Food can be considered as a national identity marker for migrants, and it produces national belonging. Interestingly, we depict something similar to Swiss markets and their customers: our results show that food markets can be analysed as a form of Swiss national identity politics. Hence, we suggest 'de-migranticising' (Dahinden, 2016) research on migration and combining it with research on food, nationalism and identity. In this way, the idea of food as an expression of national identities is not relegated solely to migrants. Instead, from a social science perspective, it would be interesting to bring together these fields of research and investigate the role of food and related institutions such as markets for national identities, be it of migrants or non-migrants. These results demonstrate that a focus on markets and the production of their supposedly local character is an interesting case to carve out different facets of a global identity politics.

Acknowledgements

We would like to thank Julia Meier, who skilfully conducted several of the interviews and observations on which this chapter builds. The project Moving Marketplaces is financially supported by the HERA Joint Research Programme (www.heranet.info). The Open Access publication of this chapter was funded by the Swiss National Science Foundation.

References

Alkon, A.H. & McCullen, C.G. (2011), Whiteness and farmers markets: Performances, perpetuations ...contestations? *Antipode*, 43, 937–959.
Antonsich, M. & Skey, M. (2017), Affective nationalism: Issues of power, agency and method. *Progress in Human Geography*, 41, 843–845.
Appadurai, A. (1990), Disjuncture and difference in the global cultural economy. In: Featherstone, M. (Ed.), *Theory, Culture & Society*, 7(2–3), 295–310. doi:10.1177/026327690007002017.
Arnaut, K., Raeymakers, T. & Schilliger, S. (2015), *New Plantations. Arbitrating 'Seasonal Migrant Labour' in Europe*. Unpublished Working Paper. Swiss Network for International Studies (SNIS). https://snis.ch/projects/new-plantations-migrant-mobility-illegality-and-racialisation-in-european-agricultural-labour-2/
Barham, E. (2003), Translating terroir: The global challenge of French AOC labeling. *Journal of Rural Studies*, 19, 127–138.
Basch, L., Glick Schiller, N. & Szanton Blanc, C. (1994), *Nations Unbound: Transnational Projects, Postcolonial Predicaments, and Deterritorialized Nation-States*. New York: Gordon and Breach.
Billig, M. (1995), *Banal Nationalism*. London: Sage.
Creswell, T. (2010), Towards a politics of mobility. *Environment and Planning D: Society and Space*, 28, 17–31.
Dahinden, J. (2012), Transnational belonging, non-ethnic forms of identification and diverse mobilities: Rethinking migrant integration? In: Messer, M., Schroeder, R. & Wodak, R. (Eds.), *Migration: Interdisciplinary Perspectives* (pp. 117–128).Vienna: Springer.
Dahinden, J. (2016), A plea for the 'de-migranticization' of research on migration and integration. *Ethnic and Racial Studies*, 39, 2207–2225.
Dahinden, J. (2017), Transnationalism reloaded: The historical trajectory of a concept. *Ethnic and Racial Studies*, 40, 1474–1485.

Dupuis, E.M. & Goodman, D. (2005), Should we go 'home' to eat?: Toward a reflexive politics of localism. *Journal of Rural Studies*, 21, 359–371.

Eriksen, S.N. (2013), Defining local food: constructing a new taxonomy—three domains of proximity. *Acta Agriculturae Scandinavica, Section B—Soil & Plant Science*, 63, 47–55.

Forney, J. (2016), Enacting Swiss cheese: About the multiple ontologies of local food. In: Le Heron, R., Campbell, H., Lewis, N. & Carolan, M. (Eds.), *Biological Economies: Experimentation and the Politics of Agrifood Frontiers*, pp. 67–81. Routledge.

Forney, J. & Häberli, I. (2016), Introducing 'seeds of change' into the food system? Localisation strategies in the Swiss dairy industry. *Sociologia Ruralis*, 56, 135–156.

Ghorashi, H. (2017), Negotiating belonging beyond rootedness: Unsettling the sedentary bias in the Dutch culturalist discourse. *Ethnic & Racial Studies*, 40, 2426–2443.

Glick Schiller, N., Basch, L. & Blanc-Zanton, C. (eds.) (1992), *Towards a Transnational Perspective on Migration: Race, Class, Ethnicity, and Nationalism Reconsidered*. New York: The New York Academy of Sciences.

Hinrichs, C.C. (2000), Embeddedness and local food systems: Notes on two types of direct agricultural market. *Journal of Rural Studies*, 16, 295–303.

Hinrichs, C.C. (2003), The practice and politics of food system localization. *Journal of Rural Studies*, 19, 33–45.

Jaberg, S. (2018), Saisonniers sind wieder da, in viel grösserer Zahl als bisher. *swissinfo. ch*, 22.11.2018.

Menet, J. (2020), *Entangled Mobilities in the Transnational Salsa Circuit: The Esperanto of the Body, Gender and Ethnicity*. London: Routledge.

MMP Team (2020), Markets… without markets? Consequences of the pandemic on markets, public spaces and social relations. *YouTube*. https://www.youtube.com/watch?v=MO4hDWVa1g8

Morasso, S. & Zittoun, T. (2014), The trajectory of food as a symbolic resource for international migrants. *Outlines. Critical Practice Studies*, 15(1), 28–48. https://doi.org/10.7146/ocps.v15i1.15828

Schappo, P. & Van Melik, R. (2017), Meeting on the marketplace: On the integrative potential of The Hague Market. *Journal of Urbanism*, 10, 318–332.

Sheller, M. & Urry, J. (2006), The new mobilities paradigm. *Environment and Planning A*, 38, 207–226.

Urry, J. (2007), *Mobilities*. Cambridge: Polity Press.

Watson, S. (2009), The magic of the marketplace: Sociality in a neglected public space. *Urban Studies*, 46, 1577–1591.

Wimmer, A. (2002), *Nationalist Exclusion and Ethnic Conflict. Shadows of Modernity*. Cambridge: University Press.

Wimmer, A. & Schiller, N.G. (2002), Methodological nationalism and beyond: Nation-state building, migration and the social sciences. *Global Networks*, 2, 301–334.

13 The multi-scalar nature of policy im/mobilities

Regulating 'local' markets in the Netherlands

Emil van Eck, Rianne van Melik and Joris Schapendonk

Introduction

On an evening in late October 2019, about a hundred market traders from the Dutch province of Overijssel gathered in a conference room to attend a public discussion organised by the national traders association (*Centrale Vereniging voor Ambulante Handel*, CVAH). Only three months earlier, the CVAH had commissioned an independent research agency to investigate the local effects of the 2006 European Union (EU)–law "Services in the International Market Directive" (2006/123/EC; hereinafter referred to as "Services Directive") on concession contracts that market traders acquire from municipalities to sell their products in publicly owned markets. The public discussion was organised to inform traders about the consequences of two Articles of the Services Directive in particular. Articles 12 and 13 exercise a genuine "public procurement principle" (Usai, 2014) that requires Member States to introduce an equal selection procedure to choose among different candidates when the number of authorisations for an economic activity is restricted due to scarcity of natural resources such as physical space.[1] Moreover, they require that the duration of the concession at issue shall be limited without mechanisms that allow for its automatic renewal (EUR-Lex, 2006). The two Articles closely connect to the neoliberal ideology underlying the Services Directive, which aims to create a uniform legal regime between the Member States for different key sectors (among which public procurements) and desires to "eliminate barriers to the movement of services enabling entrepreneurs to invest in new [M]arkets, wherever located, in the EU State" (De Minico & Viggiano, 2017: 130).

The most salient problem that directly confronts market traders is that they no longer possess the guarantee of obtaining contracts that allow them to trade on markets for an unlimited period of time; a principle that currently prevails in most of the municipal market regulations in the Netherlands. Through intensive advocacy work, members of the executive board of the CVAH have travelled throughout the whole country to convince policy actors at different levels that municipalities should provide all traders with contracts of *at least* 15 years (CVAH, 2019: 41). This highlights the importance of the multi-scalar nature of policy impacts on the nature and production of markets at the local level. It also underscores

DOI: 10.4324/9781003197058-13

the role of different actors in shaping flows of knowledge about policies and in transferring policies themselves from one scale to the other. As McCann (2011: 108) has argued, these are increasingly important aspects of the production of public spaces, yet they have not always been adequately recognised and theorised.

Building on the work of scholars who have argued that rather immobile or geographically bounded aspects of place are inherently related to social-economic, political and institutional relations stemming from elsewhere (e.g. Amin, 2008; Massey, 1991; Sheller & Urry, 2006), we illustrate in this chapter the multi-scalar nature of the institutional framework that influences the production of Dutch markets. While doing so, we move beyond 'place-based' approaches to public spaces to argue that the everyday functions of public spaces are affected by institutional relations that extend beyond their physical confines (Van Melik & Spierings, 2020). Methodologically, we make use of a mix of qualitative methods, among which a discourse analysis of relevant policy texts, semi-structured interviews with institutional stakeholders and legal experts, and participant observations of public meetings during which the Market Directive was discussed. In our analysis, we trace how the socio-geographical relations of stakeholders (e.g. traders, CVAH members, municipal actors, national politicians, EU policymakers) evolved as they were drawn together into the Services Directive case. Our fieldwork is executed within the framework of the HERA (Humanities in the European Research Area)-funded research project *Moving Marketplaces: Following the Everyday Production of Inclusive Public Space*, which focuses on the mechanisms behind the production of markets as (inclusive) public spaces.

The multi-scalar nature of marketplace regulation

In the last decade, a developing body of public space research has emerged that centres on the relationship between markets and the state. Especially scholars interested in the impacts of urban regeneration programmes on marginalised city residents have taken up this line of inquiry. With the argument raised by González and Waley (2013: 967) that the "decline of the traditional retail markets in Britain has to be contextualised within the particular trajectory of recent neoliberal urban political economy", many scholars have responded to this call by showing how markets in a variety of other countries have been converted into gentrified consumption spaces (e.g. Öz & Eder, 2012; Janssens & Sezer, 2013; Guimarães, 2018; González, 2020).

These studies have significantly contributed to our understanding of how contemporary global urban transformations influence the everyday management and regulation of markets, such as stricter rules stipulating market traders to keep the stalls and surroundings tidy or to improve "poor displays". At the same time, it is important to highlight the fact that the neoliberal ideology underpinning such restructuring projects does not unfold in a unilateral way (Peck & Tickell, 2002; Peck et al., 2009; Van Gent, 2013). By coining the concept of "actually existing neoliberalism", Brenner and Theodore (2002) have laid an alternative theoretical foundation to study the contextual *embeddedness* of neoliberal restructuring projects, "insofar as they have been produced within national, regional and local contexts defined by the legacies of inherited institutional frameworks, policy regimes, regulatory practices, and political struggles" (Brenner & Theodore, 2002: 351, original emphasis).

While policies affecting markets are often equated with global forces (González, 2020), the global imposition of neoliberalism is underscored by strong associations with multi-scalar institutional frameworks that impinge upon their pursuit and social outcomes (Van Gent, 2013). As such, whatever the significance of neoliberalism as a global phenomenon, it cannot simply be understood as a sort of "global dust cloud" waiting to settle somewhere in a more or less fixed form (Cochrane & Ward, 2012: 6). Precisely because policies are responses to particular sets of social and political conditions, they can neither simply be replicated nor have the same effects in all places to which they are transplanted. At the same time, supranational policymaking (such as formulated through the EU) cannot deliver universally applicable policy templates (Theodore & Peck, 2012).

In a relatively recent attempt to make the study of these multi-scalar institutional processes empirically applicable, a "policy mobilities conversation" (Temenos & McCann, 2013) has emerged. This research agenda fuses the theoretical approach described earlier with the long-standing study of policy transfer in political science (e.g. Dolowitz & Marsh, 1996; Stone, 1999, 2004) and the recent mobilities approach in social sciences (e.g. Hannam et al., 2006; Sheller & Urry, 2006; Cresswell, 2010). Especially since the mid-2000s, scholars engaging with the policy mobilities conversation have shifted the debate from policy transfer to policy mobilities to reject the former's tendency to adopt a literal notion of transfer in which policies are assumed to move fully formed. The focus on mobilities, instead, connotes the flows, moorings and partitioning of policies in their movement between different geographical scales.

Furthermore, the policy mobilities conservation provides the opportunity to think about policy mobilities as socially produced, open-ended practices in terms of their movement, applications and mutations (McCann, 2011). As such, it seriously interrogates the ways in which policies shape the production of public spaces *through* power relations between actors who are multi-scalarly located. Such a process-oriented, rather than place-based, approach to the study of public space (Van Melik & Spierings, 2020) seems fruitful to fully capture the ways in which the neoliberal ideology of the Services Directive has interacted with the already-existing institutional arrangements of Dutch markets.

Case study and methods

To explore how the Services Directive has been translated through practices at multiple scales, we follow Roy's (2012) suggestion to take the "middling technocrats" as object of analysis. Middling technocrats are not simply policymakers or technical experts but actors who are specialised in a specific topic and who are increasingly moving between local, national and international institutions, reshaping these accordingly (Laurie & Bondi, 2005). Representing and advocating for the rights and voices of market traders throughout the Netherlands since 1921, (board) members of the CVAH can be conceived of as such middling technocrats. The CVAH is not so much a *maker* but a multilateral *mediator* of policy (see Theodore & Peck, 2011) which establishes arenas for policy discussion and compromise among its members. In this work, the CVAH must rely on persuasion as it lacks the immediate power to enforce, or oppose, binding policies or recommendations from other levels. As a "soft power institution", the CVAH has brought together a plethora of stakeholders and policy agents by purposefully negating the harsh effects of the Services Directive on the profession and everyday practices of traders through its organisational and advocacy strategies. The strategies that operate within and against the policy apparatus of market regulations are often fleeting, sometimes persistent (see Van Eck et al., 2020) and the task of the researcher, as Roy (2012: 37) argues, is to "capture this complex terrain of complicity and resistance".

The research for this study supports the interpretation and observations of an overarching ethnographic study that aims to enlighten the place-making and mobility dynamics of market traders in the everyday production of public space. For several months between the summers of 2019 and 2020, we conducted ethnographic research by following the everyday lives of market traders in a diverse set of Dutch markets. What is important is that their practices should be framed in relation to institutional arrangements that set and control the parameters of both their settlement in place and the translocal motion between places. We have made use of multiple qualitative research methods to study the rules and regulations affecting the everyday functioning of traders and their markets. The data-gathering cycle started with participant observations of a public discussion organised by the CVAH in October 2019 as described in the introduction. In June 2020, we attended a webinar on the same topic during which the national secretary of the CVAH discussed the developments of the effects of the Services Directive. Through this practice, we have been able to identify the most important actors engaged in the Services Directive case. Accordingly, we have conducted three semi-structured interviews with five board members of the CVAH and one interview with the deputy mayor of Economic Affairs of the municipality of Bunschoten. In order to understand the effects of the Services Directive on already-existing contract systems of two different markets in the Netherlands, we have conducted four interviews with policymakers of the municipal departments of Economic Affairs. All interviews lasted between one and two hours and were audio-recorded and fully transcribed afterwards. Informal talks with market managers on both markets helped to corroborate, or provide nuance to, their statements. Supplementing the interview data, an analysis of relevant policy documents was conducted. The empirical sections below present a number of interview quotations and extracts from these policy documents.

The services directive: multi-scalar (im)mobilities of market regulations

In the Netherlands, almost all markets fall under the authority of municipal governments. They tend to have a multi-tiered structure of control, consisting of both market managers who collect market fees and directly enforce municipal regulations, and traders in market committees who resolve potential problems and issues which are reported back to the municipal council. All these regulations are included in overarching Market Statutes (*Marktverordening*). Such Market Statutes outline required behaviour and include the specificities of the selection procedure determining how traders obtain contracts for spots that have opened up on the market terrain. Different forms of selection procedures have been developed such as drawing lots, allocating by means of quality criteria or selecting on the basis of seniority (i.e., order in which traders are placed on market lists). This latter principle of issuing contracts of unlimited duration on the basis of seniority has become deeply entrenched in most of the Market Statutes in the Netherlands (interview CVAH, June 2020).

It is especially these two characteristics of the regulatory landscape of markets that the Services Directive aims to dismantle. Both characteristics—that of seniority and unlimited duration—obstruct the equal opportunity for new tenders to obtain scarce contracts and therefore undermine the underlying neoliberal ideology of free establishment and investment in markets. The EU can employ several coercive mechanisms, such as treaties, European legislation or quasi-juridical power, to shepherd Member States into desired policy directions. Directives, represent a soft mode of governance, combining both coercive and voluntary properties (Kortelainen & Rytteri, 2017). While members states have to accept directives and are obliged to obey to the policy framework set by the Council of Ministers, they have some room for manoeuvre in deciding how to do this and which regulatory instruments they wish to deploy. Kortelainen and Rytteri (2017) note that both the relatively high level of abstraction of EU directives and the autonomy that Member States have in their interpretation and enactment enable the translatability and mobility of EU directives.

The abstract nature of the 2006 Services Directive, however, initially allowed for the direct opposite: a process of slow local implementation which took almost ten years and defined, as such, its first instance of sheer *im*mobility—something which Carr (2014) has labelled as "policy paralysis". In October 2015, as a decisive moment of policy change, or "policy window" (see Kingdon, 2003), the Dutch Council of State (*Raad van State*, RvS) eventually ruled that municipal contract systems, in general, have to abide to Article 12 and 13 of the Services Directive. The court decision that acknowledged the overriding force of the Articles of the Services Directive materialised when a local service provider offering boat trips through Amsterdam's canal ring appealed the municipality for issuing boat contracts of unlimited duration after a municipal decision had earlier denied such authorisation for him (see EUR-Lex, 2015). Against the refusal, the provider pleaded the decision to the Council of State, maintaining that the policy pursued by the municipality is in conflict with the provisions on the freedom of establishment as contained in the Services Directive (Faustinelli, 2017). Asking about the possible reasons behind this relatively long period of policy immobility, a policymaker of Economic Affairs of the municipality of Amsterdam responded,

> The Services Directive hadn't been implemented in many localities during that time, especially because municipalities were hesitant to deprive traders

from their acquired rights [i.e., seniority principle]. And it was only from [approximately] 2016 onwards that [national] jurisprudence has been developed with regard to the application of the Services Directive principles. This jurisprudence of the Council of State has rendered the abstract principles of the Services Directive more concrete.

(Interview August 2020)

In April 2017, RvS confirmed again the required implementation of the Service Directive at the local level. This time, the court case directly applied to the market trade sector. A vendor selling flowers in Doorn appealed the decision of the municipality that had issued a contract of unlimited duration to another flower vendor (see RvS, 2017). Both court cases can be conceptualised as "mobilisation practices", as we would call them, which together have further shaped and crystallised legalisation at the local level. As a result, the board members and lawyer of the CVAH started to recognise that similar court cases in the future can put the existing contract systems at risk (interview CVAH, June 2020). Responding to this situation, the CVAH consulted a professor in administrative law at Leiden University who advised the CVAH to draft a report on the negative consequences of the Services Directive for traders and to propose on the basis of those arguments a suitable duration of contracts within the legal boundaries of the EU directive.

Accordingly, in November 2018, after a two-year period of "intensive advocacy" and sustained contact with the House of Representatives (CVAH, 2018), two parliamentarians, Stoffer and Wörsdörfer, filed a motion requesting the national government to "consult with the *Vereniging van Nederlandse Gemeenten* [VNG, Association of Netherlands Municipalities] in the short term" to decide upon an approach to alleviate the prevailing uncertainty among traders in the face of short-term contracts and to "ensure the survival of the ambulant trade sector" (Tweede Kamer, 2018). In order to prevent municipalities throughout the country from dissolving existing contracts with traders in the period between the adoption of the motion and the ministerial response (which had to await the research results of the CVAH report), "immobilisation strategies" were deployed to delay the immediate local implementation of the Services Directive. After consultations with the CVAH, the alderman for Economic Affairs of the municipality of Bunschoten (a municipality which might be said to constitute the "core" of the Dutch ambulant trade sector due to its history of fishery) decided to write a letter to all 354 deputy mayors of Economic Affairs throughout the country, requesting them to "not make changes in the Market Statutes until the outcomes of the national discussion" (Municipality of Bunschoten, 2019). The deputy mayor explained the underlying argumentation for this immobilisation strategy as follows:

The main problem for us is not so much our own market ... but that our entrepreneurs work on different markets across the whole country. As such, we have a completely different view on the importance of this sector than other municipalities where the ambulant trade, for example, only covers two percent of the economy. Especially such municipalities are more likely to think: "Hey, there has been a court judgement by the Council of State, we have to do something. We want to put our house in order, so we are going to adjust our Market Statutes to fall in line with legislation." What then happens is that one municipality will decide upon contracts of four years, another on two years and again another on ten years. This confronts market traders with big uncertainty; first because of the different contract regimes in different municipalities, and secondly because these contracts are too short to recoup large investment costs. If traders pull investments out of their markets, they will eventually die out. ... We decided that it would be strategic to write a letter and bring municipalities, policy makers, aldermen and councillors to the attention of this problem and to first await the research results of the CVAH.

(Interview September 2020)

Underscoring the importance of a consistent and homogeneous contract system for all traders among different municipalities, the deputy mayor further explained that the Services Directive, "with an abstract framework pursuing free market access", has the opposite effect of creating a nationally scattered institutional landscape that—according to the deputy mayor and the CVAH—constrains traders' possibilities to safely invest and settle in different municipal markets. The Services Directive takes, using the alderman's verbiage, a "one-size-fits-all-form" that "goes against local and national trade practices that have been going on for generations". He concluded, "The complexity of the functioning of markets is much bigger than pursuing more market accessibility. I'm sure everyone wants that [sic], but its implementation has inadvertent effects".

In the long-awaited ministerial response to the national motion of 2018, Mona Keijzer, former state secretary for Economic Affairs and Climate Policy, eventually decided to not incorporate the advice of the CVAH report (2019) that encourages municipalities to universally issue contracts of a minimal 15 years to all traders. Rather, Keijzer writes in her letter to the Parliament that the "stimulation and exchange of best practices among municipalities and trade unions [i.e., CVAH] can help to learn from each other" in deciding upon the "fulfilment of contract systems and the substantiation of the duration of contracts" (Tweede Kamer, 2019: 5). Obscuring the details of how such "best practices" should look like be put in place (see Bulkeley, 2006), the Ministry of Economic Affairs has opted to leave it to the municipalities how to concretely implement the Services Directive at the local level.

Recently, however, the CVAH has been notified that the Ministry of Economic Affairs and Climate Policy has consulted a new research agency to look into new possibilities to set different, fixed durations of contracts for different branches within the ambulant trade sector, while dismissing the CVAH report as a "lobby report" (interview CVAH, June 2020). The national secretary of the CVAH lamented,

This has really upset us. ... And to be honest, I'm worried. How can we, as national traders, association, justify that we agree with different contract

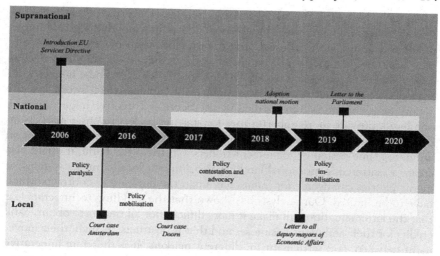

Figure 13.1 The mobilisation process of the Services Directive.
Source: authors.

durations for different branches? That is a delicate issue. I want to know how we can have a say in this and make clear that we oppose this development. We really oppose this. Because it's dangerous. Anyway, this is what is happening right now.

(Interview June 2020)

In sum, Figure 13.1 shows the timeline of the mobilisation process of the Services Directive, following a multi-scalar route that is structured around different moments and practices of im/mobility. Both Figure 13.1 and our empirical evidence make clear that multi-scalarly emplaced practices and moments of mobility-immobility do not simply represent an asymmetrical dichotomy, but rather exist in a dialectical relationship in the making and remaking of policies (Jacobs, 2011; Oancă, 2015).

Conclusion

The aim of this chapter has been to investigate the governance of markets "in action" (McCann, 2013) and to understand it as a dynamic assembly of practices that, albeit 'locally' situated, also resonates 'extra-locally' in complex and contested ways. Moving beyond the general tendency in studies on markets to explain the rules and regulations managing markets as mere derivatives of neoliberal global forces, we have shown that local policy domains are not passive in the face of these forces or logics. Rather, municipal Market Statutes regulating the management of marketplaces are actively made and re-made through the power relations between actors who are multi-scalarly located.

As such, by taking the neoliberal EU Services Directive as our empirical starting point, we have been able to show that policy mobilisation and implementation do not occur through one-to-one replication but through a "complex sociospatial process of emulation and transmutation that has uneven consequences for cities and citizens" (McCann, 2013: 20). These uneven consequences beg further research attention. As a "double-edged sword", as the deputy mayor in Bunschoten called it, the Services Directive might, one the one hand, provide more opportunities for starting traders to enter markets, as they no longer face waiting lists for market spots that are now occupied for an unlimited time period by incumbent traders. On the other hand, it might also create a scattered institutional landscape of different local contract systems, as municipalities can decide for themselves upon the duration of contracts which have to be limited. Our analysis has shown that the middling technocrats fear that the latter situation will make it more difficult for all traders to obtain bank credits for their ambulant businesses and that it obstructs, as such, their movement between, and settlement in, different markets. It is therefore imperative to approach this issue as an empirical question, opening up research venues to further unravel the effects of multi-scalar institutions on the everyday functioning of public spaces.

Note

1 In principle, the Services Directive inhibits Members States from subjecting economic services to an authorisation scheme (such as a concession contract system) *in toto*, unless contracts are scarce; a situation that applies to contracts for (physically limited) market spots. In the latter case, municipalities are only allowed to issue contracts of a limited time period and the contract system has to satisfy three cumulative conditions: non-discrimination, necessity and proportionality (EUR-Lex, 2006).

References

Amin, A. (2008), Collective culture and urban public space. *City*, 12(1), 5–24.

Brenner, N. & Theodore, N. (2002), Cities and the geographies of "actually existing neoliberalism". *Antipode*, 34(3), 349–379.

Bulkeley, H. (2006), Urban sustainability: Learning from best practice? *Environment and Planning* A, 38, 1029–1044.

Carr, C. (2014), Discourse yes, implementation maybe: An immobility and paralysis of sustainable development policy. *European Planning Studies*, 22(9), 1824–1840.

Cochrane, A. & Ward, K. (2012), Researching the geographies of policy mobilities: Confronting the methodological challenges. *Environment and Planning* A, 44, 5–12.

Cresswell, T. (2010), Towards a politics of mobility. *Environment and Planning D: Society and Space*, 28(1), 17–31.

CVAH (2018), Brede steun Tweede Kamer voor onderzoek gevolgen schaarse vergunningen ambulante handel. CVAH, 28 November 2018. Available at: https://www.cvah.nl/brede-steun-tweede-kamer-voor-onderzoek-gevolgen-schaarse-vergunningen-ambulante-handel/. Accessed on 10 September, 2020.

CVAH (2019), *Schaarse vergunningen op de markt: een onderzoek naar de gevolgen.* Capelleaan den IJssel: CVAH/GARMA BV.

De Minico, G. & Viggiano, M. (2017), Directive 2006/123/EC on services in the internal market. In: Lodder, A.V. & Murray, A.D. (Eds.), *EU Regulation of E-Commerce: A Commentary* (pp. 128–145). Northampton, MA: Edward Elgar Publishing.

Dolowitz, D. & Marsh, D. (1996), Who learns what from whom: A review of the policy transfer literature. *Political Studies*, 44, 343–357.

EUR-Lex (2006), Directive 2006/123/EC of the European Parliament and of the Council of 12 December 2006 on services in the internal market. *Official Journal of the European Union*, 27 December 2006. Available at: https://eur-lex.europa.eu/legal-content/EN/TXT/?uri=celex%3A32006L0123. Accessed on 25 May 2020.

EUR-Lex (2015), Judgement of the court (Third Chamber) R.L. Trijber v. College van burgemeester en wethouders van Amsterdam. *Portal of the Publication Office of the EU*, 1 October 2015. Available at: https://eur-lex.europa.eu/legalcontent/NL/TXT/?uri=CELEX:62014CJ0340. Accessed on 10 September 2020.

Faustinelli, E. (2017), Purely internal situations and the freedom of establishment within the context of the services directive. *Legal Issues of Economic Integration*, 44(1), 77–94.

González, S. (2020), Contested marketplaces: Retail spaces at the global urban margins. *Progress in Human Geography*, 44(5), 877–897.

González, S. & Waley, P. (2013), Traditional retail markets: The new gentrification frontier? *Antipode*, 45, 965–983.

Guimarães, P.P.C. (2018), The transformation of retail markets in Lisbon: An analysis through the lens of retail gentrification. *European Planning Studies*, 26(7), 1450–1470.

Hannam, K., Sheller, M. & Urry, J. (2006), Editorial: Mobilities, immobilities and moorings. *Mobilities*, 1, 1–22.

Jacobs, J.M. (2011), Urban geographies I: Still thinking cities relationally. *Progress in Human Geography*, 36(3), 412–422.

Janssens, F. & Sezer, C. (2013), 'Flying markets': Activating public spaces in Amsterdam. *Built Environment*, 39(2), 245–260.

Kingdon, J.W. (2003), *Agendas, Alternatives and Public Policies*. New York: Harper Collins College Publishers.

Kortelainen, J. & Rytteri, T. (2017), EU policy on the move: Mobility and domestic translation of the European Union's renewable energy policy. *Journal of Environmental Policy & Planning*, 19(4), 360–373.

Laurie, N. & Bondi, L., (Eds.) (2005), *Working the Spaces Neoliberalism*. London: Blackwell.

Massey, D. (1991), A global sense of place. *Marxism Today*, 38, 24–29.

McCann, E. (2011), Urban policy mobilities and global circuits of knowledge: Toward a research agenda. *Annals of the Association of American Geographers*, 101(1), 107–130.

McCann, E. (2013), Policy boosterism, policy mobilities, and the extrospective city. *Urban Geography*, 34(1), 5–29.

Municipality of Bunschoten (2019), College vraagt aandacht voor schaarse vergunningen ambulante handel. *Municipality of Bunschoten*, 23 May 2019. Available at: https://www.bunschoten.nl/college-vraagt-aandacht-voor-schaarse-vergunningen-ambulante-handel. Accessed 11 September 2020.

Oancă, A. (2015), Europe is not elsewhere: The mobilization of an immobile policy in the lobbying by Perm (Russia) for the European Capital of Culture title. *European Urban and Regional Studies*, 22(2), 176–190.

Öz, Ö. & Eder, M. (2012), Rendering Istanbul's periodic bazaars invisible: Reflections on urban transformation and contested Space. *International Journal of Urban and Regional Research*, 36(2), 297–314.

Peck, J. (2011), Geographies of policy: From transfer-diffusion to mobility-mutation. *Progress in Human Geography*, 35(6), 773–797.

Peck, J., Theodore, N. & Brenner, N. (2009), Neoliberal urbanism: Models, moments, mutations. *SAIS Review of International Affairs*, 29(1), 49–66.

Peck, J. & Tickell, A. (2002), Neoliberalizing space. *Antipode*, 34, 380–440.

Roy, A. (2012), Ethnographic circulations: Space-time relations in the worlds of poverty management. *Environment and Planning A*, 44, 31–41.

RvS (2017), Uitspraak 201602663/1/A3. *Raad van State*, 12 April 2017. Available at: https://www.raadvanstate.nl/@107161/201602663-1-a3/. Accessed 11 September 2020.

Sheller, M. & Urry, J. (2006), The new mobilities paradigm. *Environment and Planning A*, 38, 207–226.

Stone, D. (1999), Learning lessons and transferring policy across time, space and disciplines. *Politics*, 19(1), 51–59.

Stone, D. (2004), Transfer agents and global networks in the "transnationalization" of policy. *Journal of European Public Policy*, 11(3), 545–566.

Temenos, C. & McCann, E. (2013), Geographies of mobilities. *Geography Compass*, 7(5), 344–357.

Theodore, N. & Peck, J. (2012), Framing neoliberal urbanism: Translating 'common-sense' urban policy across the OECD zone. *European Urban and Regional Studies*, 19(1), 20–41.

Tweede Kamer (2018), Gewijzigde motie van de leden Stoffer en Wörsdörfer ter vervanging van die gedrukt onder nr. 336. *Tweede Kamer der Staten-Generaal*, 27 November 2018. Availabe at: https://www.tweedekamer.nl/kamerstukken/detail?id=2018Z21791&did = 2018D55882. Accessed 10 September 2020.

Tweede Kamer (2019), Brief van de Staatssecretaris van Economische Zaken en Klimaat, 32637-381. *Tweede Kamer der Staten-Generaal*, 5 October 2019. Available at: https://zoek.officielebekendmakingen.nl/kst-32637-381.pdf. Accessed 10 September 2020.

Usai, A. (2014), Caught between the public procurement principles and the 'Public Procurement Function' of directive 2006/123/EC. *European Procurement & Public Private Partnership Law Review*, 9(4), 228–239.

Van Eck, E.C., Van Melik, R. & Schapendonk, J. (2020), Marketplaces as public spaces in times of the COVID-19 coronavirus outbreak: First reflections. *Tijdschrift voor Economische en Sociale Geografie*, 111(3), 373–386.

Van Gent, W.P.C. (2013), Neoliberalization, Housing institutions and variegated gentrification: How the 'third wave' broke in Amsterdam. *International Journal of Urban and Regional Research*, 37(2), 503–522.

Van Melik, R. & Spierings, B. (2020), Researching public space: From place-based to process-oriented approaches and methods. In: Metha, V. & Palazzo, D. (Eds.), *Companion to Public Space* (pp. 16–26). New York: Routledge.

14 Afterword

Sara González

Introduction

This book confirms that markets are fantastic places to do research and enhance our understanding of society, in particular, cities. Researching markets is like taking a slice of a place and looking at the social and economic interconnections between people and things. If we stretch out our lens, we can trace the multi-scalar and often global retail chains that bring products to markets and we analyse changing consumption patterns and trends. If we zoom in, we can also see more micro-scale processes, such as the relationships forming between actors and the everyday life of public spaces. At a meso-level analysis, markets are embedded in particular policy regimes and political arrangements where actors interact with and within the state and economic Market principles. Although very similar could be said about so many other types of places such as homes, parks, hospitals, streets, schools or offices, markets seem to bring together a particularly complex set of connections and spheres as they are both spaces of social *and* economic exchange. The chapters in this book show us the multiple angles of looking at markets, exploring their diverse and complex nature.

This book is the latest of several contributions, some within the same series of Routledge's "Urbanism and the City" that focus on markets. Most of this research, particularly Seale and Evers (2015), Seale (2016) and my edited book (González, 2018a) have looked at markets from a relational perspective. By this, I mean that they have focused on the connections and relationships that markets create and facilitate. Seale (2016: 12), for example, sees markets as "nodes where material and intangible flows—of people, goods, time, sense, affect—come to rest, terminate, emerge, merge, mutate and/or merely pass through, and are continent and relational to one another". The work I have developed collaboratively with others has also emphasised a relational approach by making connections and comparing markets from different parts of the world (González, 2018a; González, 2020). This book, edited by Sezer and Van Melik, brings a new twist to this relational perspective, enhances this framework by emphasising three aspects: movement, representations and practices. Transcending their physical location, the editors and authors of this book look at markets as unbounded and fluid entities with research focused on process-oriented issues. How does this mobility approach enhance our understanding of markets, and more generally, place? What does it add to our

DOI: 10.4324/9781003197058-14

knowledge? In the following paragraphs, I draw on the book chapters to highlight four themes that I feel are the most important learning points from this approach.

Contested time and rhythms

One important aspect highlighted by several chapters in this book is that of time and how markets, their traders and users are involved in many different rhythms and movement patterns even during the course of one day. Memlük-Çobanoğlu and Kapusuz-Balcı (Chapter 4) develop a particular methodology based on Lefebvre's work—rhythmanalysis—to capture the "spectrum of flows, movements and relations of co-existing and encountering bodies, objects and spaces". They find that in Esat Marketplace in Ankara, different rhythms take place at the same time; some are more regular, like the daily arrival of traders and set up of stalls, others are more disruptive, like the theatrical performances of traders to attract the attention of buyers. What I find interesting is how the authors identify conflict and discordance between rhythms because the market has not been designed according to the very complex users' needs due to a top-down planning approach. Keswani (Chapter 3) also draws on the work of Lefebvre and human geographer Tim Cresswell to analyse the frantic choreographies of chai tea vendors in a marketplace in Ahmedabad, India. This research points to the fact that markets can be a challenge for planners, architects and urban designers because of their complex, changing and fluid nature; it is difficult to fit markets within the existing fixed structured of the city. In fact, Zuberec and Turner (Chapter 2) show very powerfully how, in Hanoi, the planners' and city officials' views of strict and rigid zones contrast with street traders' mobile strategies to avoid being caught in a street vending ban that the city has tried to implement since 2008.

This tension between different visions and rhythms of the city—where an official and sanitised version enters in conflict with more fluid and informal strategies and practices of traders—has existed for centuries and in fact led to the emergence of the municipal indoor market hall in the 18th century where trade could be regulated and controlled by public officials (Guàrdia & Oyón, 2015; Schmiechen & Carls, 1999). The tension persists more aggressively in cities in Africa, Asia or Latin America, where traders are at risk of eviction and losing their livelihoods if they trade in non-officially designated areas (Brown, 2017). Encouragingly, Ndaba and Landman (Chapter 5) discuss how in Warwick Junction, a busy trading area in eThekwini, South Africa, the city regulations have become more supportive of traders and street vendors over the years, which has required a more flexible way of thinking about the city. In fact, the relatively mobile nature of street trading, which the authors highlight as resilient, has meant that traders have been able to adapt the vending practices even during the difficult COVID-19 restrictions.

M/market values and land

I was struck by how powerful a mobility approach becomes when also combined with an attention to land. We might not think of land as something mobile—in fact, it is rather fixed—but land values and uses do change according to different interventions in the city by the state, private businesses and the public. Markets have historically occupied central areas of the city, as in fact, many markets were the birthplaces of the city. However, as cities have grown and land markets have become more competitive, with land becoming a financial asset, the traditional marketplace has often been displaced to more peripheral and marginal parts of cities to make way for more 'profitable' land uses. In their chapters, Mady and Venkov both discuss these issues in markets in Beirut (Lebanon) and Sofia (Bulgaria), respectively.

In Lebanon, Mady (Chapter 7) analyses Souq Al-Ahad market in a peripheral location in Beirut, where a market has emerged over the years selling low-cost goods by many of the Syrian refugees hosted in the city. Mady uses the concept of liminality to understand how the market changes meaning and value according to the different actors involved in it and the urban land dynamics. The marginal location of the market, with lower rents and less attention from the state and private investors, has facilitated the flourishing of many activities and social relations on the economic margins which would not be possible in a more central space in the city. However, as urban dynamics have changed and the land around the market has become strategic for real estate-led developments, the kind of activities that flourished in the marketplace have entered into conflict with the Market values of the land where it sits, and it is now under threat of displacement. As the Market principles of profit become more prevalent and the concept of value is equated to that of price, the marketplace, where other forms of value such as reciprocity or solidarity circulate too, is threatened. Playing with the relationship between Market principles and the marketplace (M/market) can lead to fruitful lines of exploration.

Venkov (Chapter 8) finds similar patterns of marginalisation in his analysis of open-air marketplaces in Sofia. It is fascinating to see how as Bulgaria's economy transitioned towards a free Market, the marketplaces, which were an important element of the socialist state, entered a process of decline. Venkov analyses this decline, not as a natural process of changes in consumer demand but as a collection of practices and representations by state and private actors that led to the socio-spatial sorting of marketplaces to the margins of the city. The marketplaces in Sofia suffered from a process that in housing studies is called *residualisation*. As markets declined due to various pressures, they became increasingly the place for the urban poor which in turn stigmatised these places. Even though marketplaces might have stayed in the same site, they were displaced as significant socio-economic spaces and marginalised in the collective imaginary of the city. Again, as the Market principles dominated the Bulgarian society, the marketplace was marginalised in favour of other forms of retail such as superstores or shopping centres.

In both of these cases we can see how the marketplace, even if it is constituted by material and fixed elements, acquires different values and representations depending on how it relates to the wider and dominant trends in the city. The economic, social and cultural value of markets is therefore mobile and fluid. What we see in these chapters is that the more social aspects of the marketplace, which support solidarity and reciprocity relationships, tend to be marginalised once the Market principles of profit and commodification become dominant.

The location of a marketplace might be (temporarily) fixed, but the real estate value of that location fluctuates affecting the economic chances of traders.

Mobile policy agendas

Several other chapters in the book also discuss how the role and function of markets in the city and urban policy are fluid and can change according to shifting agendas and different contexts. The COVID-19 pandemic has, of course, radically affected markets as many had to close or adapt to stringent new regulations. In this book, we learn how, in the face of the COVID-19 crisis, some public authorities are reframing markets as community resources. In Barcelona, a city internationally known for a strong network of municipal markets, Lindmäe and Madella (Chapter 6) explain how municipal privatisation and economic growth-led agendas have led to the gentrification and touristification of some of the city markets in the last decades. In particular, the Boqueria Market located at the centre of the touristified old town has become a global tourist attraction with the market shifting its focus away from serving the needs of the local population. However, in the last few years, changing municipal agendas to mitigate against touristification have made an effort to return the market to a neighbourhood resource. The COVID-19 crisis has accelerated this policy as the tourist flow has declined. However, the municipality is struggling to assert this new framing for the market as previous policies had eroded its public role.

In Lima, Huaita-Alfaro (Chapter 10) clearly shows how the COVID-19 pandemic laid bare the problems of neglect and disinvestment resulting from the policy of privatising markets that had been pursued in the previous decades. At the same time, the pandemic also revealed the strong role of markets as spaces for provisioning and fulfilling the everyday needs of many residents. Public authorities now realise the strong civic role that markets could play by bringing together various urban agendas around work, food or social inclusion.

Schappo's Chapter 9 on markets in Belo Horizonte also reflects on the shifting policy agendas. Like in the cases of Barcelona, Sofia, Lebanon and Lima, authorities have tended to erode the potential of markets as elements of social justice urban agendas. In her research, she charts the decline of public markets and their privatisation as local authorities no longer prioritise them. However, it was interesting to see how public authorities are making some efforts, and there is a plan for a social committee, where neighbourhood associations will be included to scrutinise the social role of markets.

If several chapters discuss how Market-led policies have impacted negatively on public marketplaces, a mobility approach helps us to refine our understanding of policymaking as complex, mobile and multi-scalar, which can generate diverse and contested outcomes. Van Eck et al.'s Chapter 13 does just that; it analyses the twists and turns of the European Union (EU) Services Directive, which aims to promote unrestricted trade across the EU and would make street trading in the Netherlands very precarious. The authors show how the Directive is interpreted variously by local authorities, national-level government bodies and trader organisations; policies are not fixed but politically constructed.

Research on markets can definitely benefit from interacting more closely with the policy mobilities approach. For example, I have been interested in how certain "market models" have become global recipes for redeveloping markets promoting the idea of gentrified and foodie spaces for middle-class consumption (González, 2018b). Market authorities in Barcelona, as mentioned by Lindmäe and Madella (Chapter 6) have for example embarked on international consultancy contracts to adapt some of the knowledge on markets to other cities. As Van Eck et al. note, these 'recipes' are not fixed but adapted, contested and transformed as they become embedded into local contexts, but still, they carry certain basic principles of how markets should be valued.

Mobile notions of place

Finally, Menet and Dahinden's and Watson and Breines's contributions present reflections on the mobile nature of the concept of place and its intersection with other concepts, such as identity and belonging, as they are constructed in a multi-scalar way. In England, Watson and Breines (Chapter 11) compare a market based in a rural setting with one in an inner-city area. By studying the mobility patterns of traders and their connections to the place where they trade, the chapter explores various forms of belonging, which might have seemed counterintuitive. Traders in the rural market are more mobile and have fewer attachments to their trading locales, while the urban traders, who always work from the same location, have developed long-term attachments and strong social relationships around their market. This chapter complicates the idea of markets being "local", something which Menet and Dahinden further study in their research on markets in Switzerland. In Chapter 12, they reveal how the notion of the "local" is something actively constructed by traders and market users to create a sense of authenticity and a higher moral ground; buying "local" was considered better by the customers. However, as the research shows, "local" was often bound with nationalistic concepts, and it obscured other transnational connections—for example, the labour of international migrants to produce the food that many customers labelled as local.

Towards an action research agenda

This transversal analysis of the chapters in this book has shown how a mobility approach, espoused by its authors and editors enhances our understanding of markets as intersections of people, practices and values. And as we have seen, often the market is the place where these fluid and contested elements clash. As an action-researcher, I am also interested in how we can take these research agendas forward to make or retain markets as inclusive spaces for a more just city. Schappo's chapter struck a note with this ambition when she mentioned that "marketplaces can be employed as part of a planning strategy to secure higher levels of urban equity and justice, but they do not automatically embody progressive ideas". Indeed, in themselves, markets do not necessarily constitute more equitable places than others as they are situated in complex, uneven and often unjust institutional arrangements.

Markets are extremely diverse and take very different formats. In the wealthy towns of Switzerland, market users worry about the provenance of their strawberries, and in Barcelona, market traders want to hang on to their tourist customers. In Beirut, Syrian refuge traders' livelihoods are under threat, while in Hanoi, street traders are chased by authorities and threatened with fines. There are no easy ways to make markets truly just spaces because they bring together contested notions of what should be valued and how. Going back to the m/Market dichotomy mentioned earlier, it seems to me, after reading through these chapters and from our own research of markets in the United Kingdom (González et al., 2021; Taylor et al., 2021; Waley et al., 2021), that we have to reject as much as possible the principles of the Market principles where an economic exchange is reduced to monetary price and value and where public space is commercialised. Instead, we should aim to expand many of the principles that we find in the marketplace where an economic exchange is imbued with social value and where solidarity and reciprocity exist alongside trade.

Markets, as Schappo notes in Chapter 9, are not intrinsically just, and therefore, there is work to be done to make them accessible and inclusive, particularly for disadvantaged groups. Academics and researchers can play a role, particularly by learning from traders and market users and engaging with public authorities who own, manage and/or regulate markets. A mobility approach to the study of markets can make us more attuned to the contested values that come into play at markets and how shifting policy agendas can change the value that authorities ascribe to markets, whether they are seen as a valuable social and economic resource or a public nuisance or a mere income generation activity. As academics, we can pivot our research agendas to make a case for the social justice potential of markets and the opportunities that are lost if traders and markets are not fully integrated into urban planning and policies to create a more socially and environmentally just city.

References

Brown, A. (2017), *Rebel Streets and the Informal Economy: Street Trade and the Law.* London: Routledge.

González, S. (2018a), *Contested Markets, Contested Cities: Gentrification and Urban Justice in Retail Spaces.* London: Routledge.

González, S. (2018b), La «gourmetización» de las ciudades y los mercados de abasto. Reflexiones críticas sobre el origen del proceso, su evolución e impactos sociales. *Boletín ECOS.* mADRID: FUHEM, 43, 1–8.

González, S. (2020), Contested marketplaces: Retail spaces at the global urban margins. *Progress in Human Geography*, 44(5), 877–897.

González, S., Taylor, M., Newing, A., Buckner, L. & Wilkinson, R. (2021), Grainger Market: A Community Asset at the Heart of Newcastle upon Tyne. Report available at: https://trmcommunityvalue.leeds.ac.uk/wp-content/uploads/sites/36/2021/06/210602-M4P-Grainger-FINAL.pdf

Guàrdia, M. & Oyón, J.L. (2015), Introduction: European markets as Markets of Cities. In: Guàrdia M. & Oyón J.L. (eds.), *European Markets as Markets of Cities* (pp. 11–71). Barcelona: Museu d'Història de Barcelona: Institut de Cultura: Ajuntament de Barcelona.

Schmiechen, J. & Carls, K. (1999), *The British Market Hall: A Social and Architectural History*. New York: Berghahn Books.

Seale, K. (2016), *Markets, Places, Cities*. London: Routledge.

Seale, K. & Evers, C. (2015), Informal urban street markets: International perspectives. In: Evers, C. & Seale, K. (eds.), *Informal Urban Street Markets* (pp. 1–14). London: Routledge.

Taylor, M., Watson, S., González, S., Buckner, L., Newing, A. & Wilkinson, R. (2021), Queen's Market: A Successful and Specialised Market Serving Diverse Communities in Newham and Beyond. Report available at: https://trmcommunityvalue.leeds.ac.uk/wp-content/uploads/sites/36/2021/06/210531-M4P-Queens-FINAL.pdf

Waley, P., Taylor, M., González S, Newing, A., Buckner, L. & Wilkinson, R. (2021), Bury Market: Shopping Destination and Community Hub. Report available at: https://trmcommunityvalue.leeds.ac.uk/wp-content/uploads/sites/36/2021/06/210531-M4P-Bury-FINAL.pdf

Index

Page numbers in *italics* refer figures and **bold** refer tables.

Printed in the United States
by Baker & Taylor Publisher Services